*Does Altruism Exist?*

FOUNDATIONAL QUESTIONS IN SCIENCE

At its deepest level, science becomes nearly indistinguishable from philosophy. The most fundamental scientific questions address the ultimate nature of the world. Foundational Questions in Science, jointly published by Templeton Press and Yale University Press, invites prominent scientists to ask these questions, describe our current best approaches to the answers, and tell us where such answers may lead: the new realities they point to and the further questions they compel us to ask. Intended for interested lay readers, students, and young scientists, these short volumes show how science approaches the mysteries of the world around us, and offer readers a chance to explore the implications at the profoundest and most exciting levels.

# Does Altruism Exist?

*Culture, Genes, and the*
*Welfare of Others*

David Sloan Wilson

Yale UNIVERSITY PRESS
NEW HAVEN AND LONDON

Templeton Press

Yale University Press books may be purchased in quantity for
educational, business, or promotional use. For information, please e-mail
sales.press@yale.edu (U.S. office) or sales@yaleup.co.uk (U.K. office).

Designed by Gopa & Ted2, Inc.
Set in Hoefler Text by Gopa & Ted2, Inc.

Printed in the United States of America.

Library of Congress Control Number: 2014942467
ISBN 978-0-300-18949-0 (cloth: alk. paper)

A catalogue record for this book is available from the British Library.

This paper meets the requirements of ANSI/NISO Z39.48-1992
(Permanence of Paper).

10 9 8 7 6 5 4 3 2 1

To altruists everywhere,
defined in terms of action

# Contents

# *Acknowledgments*

WRITING THIS book made me realize how much altruism has been a central question throughout my career, from my first paper on group selection as a graduate student to my current studies of human altruism in the context of everyday life. I therefore have a lot of people and organizations to thank. In the interest of brevity, Elliott Sober, Christopher Boehm, and Elinor Ostrom deserve special mention for their foundational contributions to the study of altruism. It has been an honor and privilege to work with them.

Altruism requires becoming part of something larger than oneself. I would therefore like to thank two organizations: EvoS, Binghamton University's campuswide evolutionary studies program that has grown into a multi-institution consortium, and the Evolution Institute, the first think tank that formulates public policy from a modern evolutionary perspective. I helped to create these organizations and know well how many people generously contributed their time and effort. I am especially grateful to Jerry Lieberman and Bernard Winograd for their help bringing the Evolution Institute into existence.

Among funding organizations, the John Templeton Foundation

(JTF) deserves special mention. Its mission to fund scientific research on "The Big Questions" is unique, and many of its projects would have virtually no chance of being supported by more conventional funding agencies. In addition, topics identified by JTF frequently enter the scientific mainstream after JTF has led the way. I met Sir John Templeton only once, at a workshop on the concept of purpose that he attended, and I was struck by his pragmatism in addition to his focus on the Big Questions. My focus on altruism defined in terms of action and in the context of everyday life is in the same spirit.

I have been surrounded by altruists in my own life, which makes it easy to be an altruist in return. My love to them all.

*Does Altruism Exist?*

# *Introduction*
## ALTRUISM AND EVOLUTION

ALTRUISM IS a concern for the welfare of others as an end in itself. Improving the welfare of others often requires a cost in terms of time, energy, and risk. Even the simple act of opening a door for someone requires a tiny expenditure of time and energy. At the opposite extreme, saving a life often requires a substantial risk to one's own.

Seemingly altruistic acts take place all around us, from simple courtesies to heroic self-sacrifice. The question "Does altruism exist?" might therefore seem like a silly topic for a book. Yet the claim that altruism does *not* exist has a long tradition in philosophical, political, economic, and biological thought. Those who challenge the existence of altruism do not deny that there are seemingly altruistic *acts* but question whether they are based upon altruistic *motives*. Pure egoists might help others as a means to pursue their own selfish ends, but they don't qualify as altruists because they don't care about the welfare of others as an end in itself.

The idea that altruism truly doesn't exist—that *all* people care *only* about themselves—might seem too extreme to take seriously.

Consider, however, that the word "altruism" didn't exist until 1851, when it was coined by the French philosopher Auguste Comte. If people are altruistic, then why doesn't the word (or its equivalent) have a more ancient pedigree?

The plot thickens when we consider the role that altruism plays in religious thought. About a decade ago, the John Templeton Foundation commissioned two eminent theologians and scholars of religion, Jacob Neusner and Bruce Chilton, to organize a conference on altruism in world religions. They began by asking one of their colleagues, William Scott Green, to outline a general framework for studying altruism that could be applied to particular religions by the other conference participants. Green reviewed definitions of altruism and provided this concise version: "Intentional action ultimately for the welfare of others that entails at least the possibility of either no benefit or a loss to the actor." According to the conference participants—each an expert scholar on a given religion—this concept is foreign to the imagination of *all* of the world's major religious traditions.[1]

I will return to Comte and religious belief systems in chapter 6. For now, suffice it to say that the question "Does altruism exist?" is not trivial and takes us into deep intellectual waters. In this book I use evolutionary theory as our navigational guide. Altruism occupies center stage in Darwinian thought because it appears difficult to explain as a product of natural selection. If natural selection favors traits that cause individuals to survive and reproduce better than other individuals, and if altruistic acts increase the survival and reproduction of others at a cost to the altruist, then how can altruistic traits evolve? This question has been debated from Darwin to the present and might seem as unresolved among evolutionists as among philosophers.[2] On the contrary, I believe that a resolution is at hand, which I attempt to convey in this book.

This book initiates a series of short books on big questions organized by the Templeton Press and Yale University Press. The idea of a short book on a big question is both enabling and constraining. It must stick to the basics and provide an accessible introduction to a large and diverse audience. It should act as a portal to the academic literature without getting bogged down in the details.[3] Ideally, it should offer something to the expert in addition to the novice. I welcome the opportunity to write a short book on the big question of altruism for three reasons.

First, some of the biggest controversies in science end up appearing obvious in retrospect, even though decades were required for their resolution. Examples include the Copernican view of the solar system, Darwin's theory of evolution, and the theory of continental drift. We find it hard to understand in hindsight why smart people took so long to reach a consensus on these bodies of knowledge. I believe that the question of how altruism evolves is such a controversy that is just entering its resolution phase. Despite a half century of disagreement, people will look back and wonder what all the fuss was about. This book operates from a postresolution perspective, which enables me to explain how altruism evolves in a short space.

Second, most of the thinking on human altruism is not from an evolutionary perspective. Foundational ideas were established by thinkers who lived before Darwin or by contemporaries who did not fully absorb the implications of his theory. Even the modern academic literature is largely not from an evolutionary perspective, for reasons that I will recount as we go along. This literature often fails to carve the subject of altruism at the right joints, to use a venerable philosophical phrase. In other words, it often fails to make distinctions that are foundational from an evolutionary perspective, such as between proximate and ultimate causation.

When we base our analysis on evolution, some questions that appeared central become peripheral and other previously ignored questions become central. As an analogy, suppose that someone owes you money and offers to pay by cash or check. You might have a mild preference for one payment method, but your main concern is to be paid. As we shall see, some criteria for defining human altruism that have occupied center stage are like payment methods. They might exist, but we shouldn't care that much about them compared to other ways of thinking and feeling that produce equivalent results. More generally, I look forward to showing how evolutionary theory can carve the topic of altruism at the right joints, or at least *different* joints, in a short space.

Third, understanding altruism from an evolutionary perspective has enormous practical import in addition to its intellectual charm. My own inquiries took a practical turn in 2006 when I began to study altruism in the context of everyday life in my hometown of Binghamton, New York. A year later I helped to create the Evolution Institute, the first think tank to formulate public policy from an evolutionary perspective.[4] These endeavors have enabled me to explore topic areas such as economics and business, where the prevailing assumption is that people are motivated entirely by self-interest. Practices that follow from this assumption have had ruinous consequences, as we shall see. Yet well-meaning efforts to promote altruism can also have pathological consequences when they do not take basic evolutionary considerations into account, as strange as that might seem. By the end of this short book, if I have done my job well, the reader will understand why evolutionary theory is essential for accomplishing the altruistic goal of making the world a better place.

## CHAPTER 1

# Groups That Work

IMAGINE THE purest of altruists, who thinks only about making the world a better place and cares nothing for her own welfare, other than as part of the common good. Now imagine convincing this person that the world of her dreams is possible. There is only one catch. The people capable of achieving and maintaining her utopia do not necessarily think or feel about themselves and others in the same way that she does. She might need to learn how to think and feel differently to accomplish her altruistic goals.

Would a pure altruist accept this bargain? It seems to me that she must. If she insisted on clinging to her habits of thought, she would interfere with her own altruistic impulse to make the world a better place.

This thought experiment highlights two meanings of altruism that need to be distinguished from each other. The first meaning refers to how people *act*. If our pure altruist were asked to describe her perfect world, she would spend more time on what people do than how they think or feel. At a local scale, they would refrain from robbery, rape, and murder. A helping hand would be offered to those who fall upon hard times. At a global scale, there would be an absence of warfare. The division between rich and poor nations would disappear. The great mass of humanity would somehow manage its affairs to avoid polluting and overheating the earth. It is impossible to talk about making the world a better place without referring to *actions* of this sort.

The second meaning of altruism refers to the *thoughts and feelings* that cause people to act as they do. Some thoughts and feelings are oriented toward the welfare of others and society as a whole, while others are more self-oriented. It might seem that the two meanings are straightforwardly related to each other. Get people to think and feel more altruistically, and the world will become a better place. If everyone could become as pure as our pure altruist, utopia can be achieved.

A little reflection reveals a more complicated story. Any given action can be motivated by a diversity of thoughts and feelings. I might help you because (a) I think it's the right thing, (b) I take pleasure in your pleasure, (c) I regard it as my ticket to heaven, (d) I am trying to improve my reputation, (e) I'm trying to put you in my debt, or (f) I'm being paid to do it. It's not obvious which of these thoughts and feelings count as most altruistic (a vs. b, for example) or if the most altruistic thoughts and feelings result in the most helping. Perhaps reputational mechanisms (d) work better than voluntary altruism (a–c), for example. Or perhaps market forces driven by self-interest result in societies that work best, as some economists contend.

These complications force us to recognize a one-to-many relationship between any given action and the mental events that can cause the action. They also force us to recognize that *our preference for some thoughts and feelings over others is based primarily on the actions they produce*. There is no other reason to privilege thoughts and feelings that count as altruistic over those that count as selfish.

In this chapter, we begin answering the question "Does altruism exist?" by investigating altruism at the level of action, deferring altruism at the level of thoughts and feelings until later. When Ted benefits Martha at a cost to himself, that's altruistic, regardless of how he thinks or feels about it. Altruism at the level of action is closely related to group-level functional organization. The reason that people need to provide services for each other is because we are a social species and mutual aid is required to accomplish together what cannot be accomplished alone. The list of activities that required mutual aid before the advent of agriculture included child care, hunting and gathering, defense against predators, and defense and offense against other human groups. The advent of agriculture led to an autocatalytic cycle between the production of resources and larger societies with division of labor, leading to modern megasocieties. Today we are so dependent upon the actions of others that we could no more survive on our own than an ant separated from its colony.

Something is functionally organized when its parts work together in a coordinated fashion to achieve a given end. The organelles of a cell and the organs of a multicellular organism are miracles of functional organization designed by natural selection to enhance survival and reproduction. The parts of a can opener are functionally organized to open cans. They are not functionally organized for chopping wood. Anyone can make this determination by attempting to use a can opener for both tasks.

When a group of organisms is functionally organized, its members coordinate their activities for a common purpose, just like the organs of an organism and the parts of a can opener. A group that is extremely well organized could even be called a *superorganism*, a metaphor with a venerable history in philosophical, religious, and political thought.

Do human groups ever achieve a degree of functional organization that deserves the term superorganism? How about animal groups? In short, do functionally organized groups exist? This is not exactly the same question as "Does altruism exist?" but it is highly relevant and a lot easier to answer, as we shall see.

To begin, consider an aircraft carrier. The ship itself is a marvel of functional design for the purpose of providing a mobile airport. Nobody who studies its features can doubt its function any more than doubting the function of the heart or the toothed wheel of a can opener. The social organization of the crew is also a marvel of functional design, as anthropologist and cognitive psychologist Edwin Hutchins describes in his book *Cognition in the Wild*.[1] Several hundred in number, the crew is organized into teams that are specialized to perform various tasks. Hutchins focuses on the team responsible for the crucial task of navigation and especially locating the position of the ship when it is close to shore. The process of measuring the compass directions of landmarks on the shore, marking them on a chart, and using triangulation to identify the position of the ship is easy to understand and can be performed by a single person given enough time, but it must be performed by a team of people to ensure accuracy and to recalculate the ship's position every few minutes. The social organization of the navigational team has been refined over the course of nautical history and has even coevolved with the physical architecture of the chartroom. It includes the training of team members over the long term and enough redundancy so that some members of the team can be called away without compromising the performance of the group too severely. One reason that Hutchins chose to study navigational teams is that their social interactions can be understood in purely computational terms—the perception and transformation of information, leading to physical action—in just the same way that a single brain can be understood in computational terms. This particular group of people engaged in this particular task is functioning like a single brain. I say more about this comparison as we proceed.

Another example of group-level functional organization serves us throughout the book. Elinor Ostrom was awarded the Nobel Prize in economics in 2009 for showing that groups of people are capable of managing their own resources, but only if they possess certain design features.[2] She and her associates did this in part by assembling a worldwide database of groups that attempt to manage resources such as irrigation systems, ground water, fisheries, forests, and pastures. These are called *common-pool resources* (CPR) because they cannot easily be privatized. They are vulnerable to a social dilemma that ecologist Garrett Hardin made famous in his 1968 article titled "The Tragedy of the Commons."[3] Hardin imagined a village with a field upon which any townsperson could graze his or her cows. The field can only support so many cows, but each townsperson has an incentive to exceed this capacity by adding more of his or her own cows, resulting in the tragedy of overuse. Prior to Ostrom's work, economic wisdom held that the only solution to this problem was to privatize the resource (if possible) or to apply top-down regulation. Ostrom's demonstration that CPR groups are capable of managing their own affairs was so new and important that it warranted the Nobel Prize.

Ostrom's work was equally noteworthy for showing that groups must possess certain design principles to effectively manage their resources. An efficacious group needs these design principles in the same way that we need our organs, a can opener needs its parts, and the navigation team aboard an aircraft carrier needs to be structured as it is. Ostrom identified eight core design principles required for the CPR groups in her worldwide database to effectively manage their affairs:

1. *Strong group identity and understanding of purpose.* The identity of the group, the boundaries of the shared resource, and the need to manage the resource must be clearly delineated.

2. *Proportional equivalence between benefits and costs.* Members of the group must negotiate a system that rewards members for their contributions. High status or other disproportionate benefits must be earned. Unfair inequality poisons collective efforts.

3. *Collective-choice arrangements.* People hate being told what to do but will work hard for group goals to which they have agreed. Decisionmaking should be by consensus or another process that group members recognize as fair.

4. *Monitoring.* A commons is inherently vulnerable to free-riding and active exploitation. Unless these undermining strategies can be detected at relatively low cost by norm-abiding members of the group, the tragedy of the commons will occur.

5. *Graduated sanctions.* Transgressions need not require heavy-handed punishment, at least initially. Often gossip or a gentle reminder is sufficient, but more severe forms of punishment must also be waiting in the wings for use when necessary.

6. *Conflict resolution mechanisms.* It must be possible to resolve conflicts quickly and in ways that group members perceive as fair.

7. *Minimal recognition of rights to organize.* Groups must have the authority to conduct their own affairs. Externally imposed rules are unlikely to be adapted to local circumstances and violate principle 3.

8. *For groups that are part of larger social systems, there must be appropriate coordination among relevant groups.* Every sphere of activity has an optimal scale. Large-scale governance requires finding the

optimal scale for each sphere of activity and appropriately coordinating the activities, a concept called *polycentric governance*.[4]

These core design principles were described in Ostrom's most influential work, *Governing the Commons*, published in 1990. A recent review of ninety studies that have accumulated since the original study provides strong empirical support for the efficacy of the core design principles, with a few suggested modifications that differentiate between the resource system and those authorized to use it.[5] I was privileged to work with Lin (as Ostrom encouraged everyone to call her) and her postdoctoral associate, Michael Cox, for several years prior to her death in 2012, resulting in an article entitled "Generalizing the Core Design Principles for the Efficacy of Groups," which I draw upon throughout this book.[6]

These two examples establish a point that I hope will be obvious to most readers: at least some human groups are impressively functionally organized, inviting comparison to a single organism. How about animal groups? Nonhuman species afford many examples of group-level functional organization, including but not restricted to social insect colonies such as ants, bees, wasps, and termites.[7] These colonies have been celebrated for their groupishness since long before science existed as a cultural institution. Beehives are pictured on the road signs in Utah because the Mormons, following a long Christian tradition, admire the industry of individual bees on behalf of their groups.

Worker honeybees forgo reproduction and are willing to die in defense of their colony. These acts appear highly altruistic in behavioral terms (although the mental world of bees is no more scrutable than that of humans), but I focus here on a more subtle aspect of colony-level functional organization: the ability to make collective decisions. Bee colonies make daily decisions on where to forage, how to allocate the worker force, and so on. In addition, when a colony splits by swarming, the swarm is faced with the critical task of finding a new nest cavity. Social insect biologist Tom Seeley and his colleagues have studied this process in minute detail.[8] For some of their experiments, they transport beehives to an island off the coast of Maine that lacks natural nest cavities. The researchers present the swarms with artificial nest cavities that differ in their size, height, distance from the swarm, compass orientation, exposure to the sun, size of the opening, physical position of the opening, and other parameters. In this fashion, the researchers have determined that honeybee swarms give the nest cavities a thorough inspection and reliably pick the best over the worst, just like a discerning human house hunter. How do they do it?

Seeley and his colleagues can answer this question because all of the scout bees have been marked as individuals and monitored on their travels from the nest cavities to the swarm. Most scouts do not visit more than one nest cavity and therefore do not have enough information to make an individual choice. Instead, the collective decision is made on the basis of social interactions that take place on the surface of the swarm. The bees that make up the swarm, including the queen, take no part in the decisionmaking process, other than providing a surface for the scouts to interact with each other. The interactions can be regarded as a contest among the scouts that is decided on the basis of their persistence, which in turn is decided by the quality of the cavity that has been inspected. Once the winning faction reaches a critical threshhold, a new phase begins that involves arousing the entire swarm and guiding it to its new home. Even to a veteran evolutionist such as myself, it seems miraculous that such a coordinated social process among mere insects could have evolved by natural selection. But then, you and I are nothing more than a coordinated social process among mere cells!

How does the decisionmaking process of a honeybee colony compare to that of an individual such as you or me? Detailed research on rhesus monkeys involves having them watch a number of dots moving either left or right on a screen.[9] The monkeys are trained to move their head in the direction that most of the dots are moving, by rewarding them with a few drops of sweet liquid if they succeed and few drops of salty liquid if they fail. As they are making their decision, the researchers record the activity of neurons in their brains. Remarkably, the interactions among the neurons resemble the interactions among the bees. Some neurons fire at the sight of right-moving dots and others fire at the sight of left-moving dots. Their firing rates are unequal because more dots are moving in one direction than the other. As soon as the stronger "faction" reaches a threshold, the decision is made and the monkey turns its head in the appropriate direction. Once we regard an individual organism as a population of socially interacting cells, the possibility of a group mind, comparable to an individual mind, becomes less strange.

Biologists expect social insect colonies to have group minds because they function as superorganisms in so many other respects. The phenomenon of group-level cognition in nonhuman species is more general, however. Adult female African buffaloes (but not juveniles or males) "vote" on where to graze at night by standing up and pointing their heads in their preferred direction with a distinctive posture while the herd is resting in the afternoon. This information is somehow integrated and, when evening falls, the entire herd gets up and moves in the chosen direction "as if conducted by a single master."[10]

I have studied group cognition in American toad tadpoles with my former graduate student Charles Sontag.[11] We created food patches that varied in quality by mixing different amounts of aquarium fish flakes, ground to a fine powder, in an agar matrix. We placed the food patches in the tadpoles' natural environment and videotaped their response. Within minutes, the tadpoles started streaming toward the food patches and discriminated the better from the worse patch. They even formed columns that bore an eerie resemblance to ant columns. In laboratory experiments, we determined that the tadpoles communicate through surface waves and that individual tadpoles appear unable to locate a food patch by themselves, even from a short distance.

In addition to empirical examples such as these, a growing body of theory suggests that group-level cognition (along with other group-level adaptations) should be found in many social species, not just social insect colonies.[12] In addition, animal groups can be functionally organized in some respects but not others. They might cooperate to locate the best food patches but compete once they get there, for example.

Let's take stock of the progress that we have made in this chapter. First, we have made a crucial distinction between altruism at the level of *action* and altruism at the level of *thoughts and feelings*. Second, we have established that altruism at the level of thoughts and feelings is worth wanting only insofar as it leads to actions that make the world a better place. Third, we have established that making the world a better place requires *group-level functional organization*—people coordinating their activities in just the right way to achieve a common goal. Fourth, we have established that group-level functional organization exists in both human and non-human societies. At least some of the time, members of groups coordinate their activities so well that they invite comparison to the cells and organs of a single organism.

The examples of group-level functional organization that I have provided do not seem very altruistic in the conventional sense. A navigational team charting the position of a ship is different from a captain going down with the ship or a soldier falling upon a grenade to protect his comrades. A group of scout bees deciding upon the best tree cavity seems different from a worker bee delivering a suicidal sting to an intruder. The key difference involves the cost of providing services to other members of one's group. The greater the cost, the more altruistic the action appears. But this is a matter of degree, not kind, as we shall see in the next chapter.

I have purposely established the major points in this chapter with minimal reference to evolution. Evolution isn't needed to make the distinction between altruism based on action and altruism based on thoughts and feelings. It isn't needed to assert the one-to-many relationship between any given action and the mental events that can cause the action. It isn't needed to assert that our preference for thoughts and actions should be based on what they cause people to do. Evolution isn't even needed to establish the functional organization of groups, any more than it is needed to establish the functional organization of a can opener. The added value of an explicit evolutionary perspective becomes apparent as we proceed.

CHAPTER *2*

# *How Altruism Evolves*

A N EVOLUTIONARY STORY is required to explain how an animal group such as a bee colony becomes functionally organized, complete with a group mind and kamikaze workers. An evolutionary story is also required to explain how human groups become functionally organized. The human story is more complicated because it includes cultural evolution in addition to genetic evolution. Nevertheless, both stories rely upon a few foundational principles that evolutionists widely accept and that can easily be explained.

The first foundational principle is that *natural selection is based on relative fitness*. It doesn't matter how well an organism survives and reproduces, only that it does so better than other organisms in the evolving population. All knowledgeable evolutionists would agree with this statement, but vast numbers of people who are not familiar with evolutionary theory are not accustomed to reasoning in relative terms. To pick one influential example, rational choice theory in economics assumes that people strive to increase utilities, defined as things needed or wanted and usually conceptualized as wealth. The assumption is that people strive to maximize their *absolute* wealth, as if they want to be as rich as possible without caring about their wealth relative to other people. Economic theory would be very different if it were formulated in terms of relative wealth, as I discuss in more detail in chapter 7.[1]

Even if you're not an economist, you might be accustomed to reasoning in absolute rather than relative terms. If option A delivers a higher payoff than option B, then shouldn't you choose option A? Not necessarily, because your best choice might depend upon your payoff *relative to other relevant individuals*. Let's say that you're playing Monopoly and I offer you one thousand dollars, subject to the constraint that every other player gets two thousand. If you accept my offer, you will be richer in absolute terms but poorer in relative terms, and it is your relative wealth that counts in the game at hand! The difference between thinking in relative vs. absolute terms is a good example of how evolutionary theory carves a subject at the right joints, or at least locates *different* joints from those found in other bodies of thought.

The second foundational principle is that *behaving for the good of the group typically does not maximize relative fitness within the group*. This is a basic matter of tradeoffs. Just as a can opener can't chop wood and a fish dies out of water, behaviors that are designed to maximize an individual's relative advantage within a group are typically different from the behaviors designed to maximize the welfare of the group as a whole. The word "typically" admits exceptions, but there can be little doubt about the rule.

To appreciate the import of this statement, imagine a group consisting of two types of individuals. Type A provides a benefit to everyone in its group, including itself, at no cost to itself. The idea of a no-cost public good might seem unrealistic but is useful for illustrative purposes. Type S benefits from type A but does not produce the public good. The two types exist in a certain proportion—let's say 20 percent A and 80 percent S. Everyone survives and reproduces better thanks to the presence of the A-types, but the *proportion* of A-types does not change. The trait is neutral as far as natural selection within the group is concerned, because it does not produce fitness *differences*.

Now let's say that there is a tiny cost associated with providing the public good. A-types deliver a benefit of 1.0 to everyone, including themselves, at a cost of 0.01 to themselves. Their net benefit is 0.99, compared to the benefit of 1.0 to S-types. They have a relative fitness disadvantage within their group, even though they have increased their absolute fitness. The proportion of A-types during the next generation will be less than 20 percent and will decrease further with every generation until the A-types go extinct. If we increase the cost of providing the public good to include cases that are more strongly altruistic, then the A-types will go extinct even faster.

In short, *the evolution of group-level functional organization cannot be explained on the basis of natural selection operating within groups. On the contrary, natural selection operating within groups tends to undermine group-level functional organization.* This statement holds not only for traits that appear overtly altruistic, such as the suicidal sting of a honeybee worker, but also for the low-cost coordination of behaviors for the good of the group, such as deciding upon the best nest cavity. The statement is so important that E. O. Wilson called it "the central theoretical problem of sociobiology" in his 1975 book *Sociobiology*.

How does group-level functional organization evolve, if not by natural selection within groups? The third foundational principle provides the answer: *Group-level functional organization evolves primarily by natural selection between groups*. Returning to our example of a no-cost public good, imagine several groups instead of just one. The proportion of A-types in the total population is 20 percent, but there is some variation among groups. Groups with more A-types will contribute more to the gene pool than groups with fewer A-types, even if the proportions of A-types don't change within groups. The existence of multiple groups and variation among groups provides the fitness *differences* required for natural selection to favor A-types over S-types.

What happens if there is a cost associated with providing the public good? In this case, natural selection within and between groups pull in opposite directions. The fitness differences among individuals within groups weigh in favor of S-types, while the fitness differences among groups in the multigroup population weigh in favor of A-types. What evolves in the total population depends on the relative strength of these opposing forces. The tug-of-war can result in one type going extinct or can maintain a mixture of both types in the population.

Now let's apply some labels to the two levels of selection. Let's call a behavior *selfish* when it increases relative fitness within groups and *altruistic* when it increases the fitness of the group but places the individual at a relative fitness disadvantage within the group. This definition is based entirely on action and not on thoughts and feelings. Also, evolutionary biologists have other definitions of selfishness and altruism based on action, which I discuss in the next chapter. For now, we can note that our definitions are perfectly sensible, especially given the first foundational principle that natural selection is based on relative fitness. Given our definitions, we can conclude that altruism evolves whenever between-group selection prevails over within-group selection. Just as Rabbi Hillel was able to state the meaning of the Torah while standing on one foot ("What is hateful to you, do not do to your neighbor: that is the whole Torah while the rest is commentary;

go and learn it."), E. O. Wilson and I provided this one-foot summary of sociobiology in a 2007 article: "Selfishness beats altruism within groups. Altruistic groups beat selfish groups. Everything else is commentary."[2]

Some examples of within- and between-group selection in the real world will bring our abstract discussion to life. Water striders are insects that are adapted to skate over the surface of water where they prey and scavenge upon other insects. Their individual-level adaptations are breathtaking, as I describe at greater length elsewhere.[3] Their bodies are held above the water by the front and back pair of legs while the middle pair acts like a pair of oars for locomotion. Their feet are so hydrophobic that sixteen water striders could be stacked on top of each other without the bottom one breaking through the water surface. In addition to serving the function of flotation and locomotion, their legs are organs of perception, sensing ripples on the water surface in the same way that our ears sense sound waves. These are individual-level traits because they influence the fitness of the individuals bearing the traits without influencing the fitness of other individuals. A water strider with hydrophobic feet survives and reproduces better than one without hydrophobic feet, no matter how they are grouped.

In addition to their individual-level adaptations, water striders interact socially and even communicate with each other by creating ripples on the water surface. Males vary greatly in their aggressiveness toward females. Some act as rapists in human terms, attempting to mate with any female without regard to her receptivity. Others act as gentlemen in human terms, mating only when approached by females. How are these individual differences maintained in water strider populations?

Omar Eldakar, one of my former PhD students, addressed

this question by creating groups of water striders consisting of six females and six males, in which the composition of males was varied from all rapists to all gentlemen and various mixes in between.[4] Within each group containing both types, the rapists outcompeted the gentlemen for mates. If within-group selection were the only evolutionary force, the gentlemen would quickly go extinct. However, rapists prevented females from feeding and therefore caused them to lay fewer eggs. This effect was so large that females in groups with all gentlemen laid over twice as many eggs as females in groups with all rapists.

The connection between this example and our theoretical discussion should be clear. Selfishness (= rapists) beats altruism (= gentlemen) within groups. Altruistic groups (= gentlemen that permit females to feed) beat selfish groups (= rapists that prevent females from feeding). The outcome depends on the balance between these opposing levels of selection. Continuing the example, suppose that Omar had composed the groups so that all of them had the same frequency of rapists and gentlemen. There would be no variation among groups to oppose within-group selection, so rapists would evolve and gentlemen would quickly go extinct. Alternatively, suppose that Omar had composed the groups so that any given group had 100 percent rapists or 100 percent gentlemen. There would be no variation within groups and the gentlemen would quickly evolve. These two cases bracket the extremes of variation within and among groups. In real multigroup populations, the partitioning of variation within and among groups falls between these extremes. What are the causes of within- and between-group variation in water striders?

Omar investigated this question by conducting a new set of experiments where the water striders were allowed to move between groups.[5] Females who entered groups containing rapists quickly left. The rapists were free to follow, but the net effect of all the movements was a considerable degree of clustering of the females around the gentlemen. Free movement creates enough variation among groups to maintain gentlemen in the population, but rapists are also maintained in the population by their within-group advantage. The tug-of-war between levels of selection results in a mix of altruistic and selfish behavioral strategies in the population, rather than one prevailing entirely over the other.

My next example involves a very different species, a type of virus called phage that preys upon bacteria. Viruses reproduce by subverting the cellular machinery of their hosts to make more viruses rather than making more host cells. A single bacterial cell invaded by a single virus can release thousands of virus progeny into the environment within hours, ready to repeat the cycle. With this explosive reproductive capacity, a phage population can wipe out its bacterial host population and thereby drive itself extinct—a biological example of Garrett Hardin's tragedy of the commons described in the previous chapter. A more "prudent" phage population would curtail its reproductive rate, but can this kind of population-level survival strategy evolve by natural selection?

Benjamin Kerr and his colleagues conducted an elegant experiment to address this question, using the T4 coliphage virus and its bacterial host E. coli (*Escherichia coli*).[6] They created a multigroup population by culturing the bacteria and phage in multiwell plates used in automated chemical analysis. A single plate contains ninety-six wells, and each well contains .2 ml of culture medium. Dispersal between wells was accomplished by transferring small amounts of medium from one well to the other with robotic pipettes, allowing the amount of dispersal to be manipulated as part of the experiment. Although a single plate containing ninety-six groups can be held in the human hand, it contains millions of bacteria and phage and persists for hundreds of generations during the course of the experiment. What happened?

Evolution has no foresight. Phage strains that differ in their reproductive capacity arise by mutation. The more prolific strains have the relative fitness advantage within single wells and replace the less prolific strains over the course of multiple generations. If they are too prolific, they drive the local E. coli population, and therefore themselves, extinct. The vacant wells are available for recolonization by bacteria and phage from other wells. Wells with less prolific strains have an advantage in this between-group process, because they persist longer and maintain themselves at higher population densities over the long term.

As with our water strider example, the connection to our theoretical discussion should be clear. Selfishness (= prolific virus strains) beats altruism (= prudent virus strains) within groups. Altruistic groups (= prudent strains that do not drive their resource extinct) beat selfish groups (= prolific strains that drive their resource, and therefore themselves, extinct). The outcome depends upon the balance between opposing levels of selection. Continuing the example, imagine eliminating dispersal among the groups. More prolific strains would eventually arise by mutation within each well, replacing the less prolific strains, and there would be no between-group process to oppose their evolution. Now imagine creating so much dispersal that the entire multigroup population becomes a single well-mixed population. Once again, the most prolific strains would prevail on the basis of their relative fitness advantage. Intermediate levels of dispersal are required to create variation in the frequency of viral strains among groups, which is a prerequisite for a process of between-group selection that can oppose within-group selection. In their elegant experiment, Kerr and his associates showed that intermediate dispersal rates created enough variation among groups to maintain less prolific viral strains in the multigroup population.

Note that nothing is being maximized in either of my two examples. Within-group selection by itself would favor highly aggressive male striders and highly prolific phage strains. Between-group selection by itself would favor highly docile male striders and phage strains that allow their bacterial hosts to persist indefinitely. One level of selection could theoretically prevail entirely over another level, resulting in a uniform population. In these two examples, however, the two levels of selection maintain individual differences in the population.

In addition to highlighting their similarities, also note the differences between the two examples. For male aggressiveness toward females in water striders, the salient groups last a fraction of a generation and variation among groups is caused by the conditional movement of individuals. For the reproductive capacity of phage viruses, the salient groups last many generations, dispersal is passive, and variation among groups is caused by colonization and extinction events. Remarkably, the same succinct summary—"Selfishness beats altruism within groups. Altruistic groups beat selfish groups. Everything else is commentary."—can be used to make sense of such different examples. Moreover, the same summary can make sense of nearly all examples of group-level functional organization in all species. I invite the reader to choose any example of a trait that contributes to group-level functional organization and perform the simple comparison of relative fitnesses within and among groups to assess the role of between-group selection in the evolution of the trait.

So far, I have treated the words "individual" and "group" as separate, along with almost everyone else who uses these words. In the 1970s, a revolutionary claim by cell biologist Lynn Margulis challenged the distinction at its core.[7] Margulis proposed that nucleated cells (called *eukaryotic*) did not evolve by mutational steps from bacterial cells (called *prokaryotic*), but were originally symbiotic associations of bacteria that became so functionally integrated that they became higher-level organisms in their own right. Evolutionary theorists John Maynard Smith and Eors Szathmary generalized this concept to explain other transitions from groups *of* organisms to groups *as* organisms, including the first bacterial cells, multicellular organisms, social insect colonies, human evolution (the subject of chapter 4), and possibly even the origin of life itself as groups of functionally organized molecular interactions.[8]

The explanation of how these major evolutionary transitions occur falls squarely within the theory that I have outlined in this chapter, with a single additional twist: the balance between levels of selection is not static but can itself evolve. In rare cases, mechanisms evolve that suppress the potential for disruptive selection within groups, causing between-group selection to become the primary evolutionary force. These groups become so functionally organized (at least for the particular traits under consideration) that they qualify as higher-level organisms in their own right.

The word *organism* in everyday language implies a high degree of functional organization. If the parts of an organism didn't work together for their collective survival and reproduction, we wouldn't call them organs. For millennia, philosophers, religious sages, and intellectuals of all stripes have metaphorically compared human societies to single organisms in their visions of an ideal society if not in reality. What's new is that the comparison is no longer metaphorical. There is a single theory of functional organization that can be applied to all levels of a multitier hierarchy of units. Any level can become functionally organized, to the extent that selection operates at that level. Lower-level selection tends to undermine higher-level functional organization. Higher-level selection causes lower-level entities to become organlike. When selection is concentrated at a given level, units become so functionally organized that we call them organisms. A multicellular organism is literally a group of groups of groups, whose members led more fractious lives in the far distant past. Moreover, even for entities that serve as the archetypes for the concept of an organism, lower-level selection is only highly suppressed and never entirely eliminated. Some lower-level entities within our bodies, such as cancer cells and genes that bias their transmission through our gametes, are the equivalent of rapist water striders and prolific viral strains.[9] If a present or future human society became so functionally organized that its members acted almost entirely for the common good, that society would deserve the term *organism* just as much as a multicellular organism such as you or me.

The fact that the organisms of today were the groups of past ages was beyond Darwin's imagination. It wasn't proposed for any biological entity until the 1970s and didn't become generalized until the 1990s. Thinking of human genetic and cultural evolution as a series of major transitions is newer still, as I explain in chapter 4. These foundational developments in evolutionary science cannot help but reorganize the vast literature on the nature of human individuals in relation to their societies that has accumulated over the course of recorded history, including but not restricted to the concept of altruism.

With respect to the core question of this book—"Does altruism exist?"—we can say this: When altruism is defined in terms of action and in terms of relative fitness within and between groups, it exists wherever there is group-level functional organization.

## CHAPTER 3

# *Equivalence*

CHARLES DARWIN appreciated the fundamental problem
of social life that makes it difficult to explain how altruism
and other morally praiseworthy traits evolve on the basis
of selection among individuals within groups. He also correctly
grasped the solution in this iconic passage from *The Descent of
Man*:

> It must not be forgotten that although a high standard
> of morality gives but a slight or no advantage to each
> individual man and his children over other men of the
> same tribe, yet that an increase in the number of well-
> endowed men and an advancement in the standard of
> morality will certainly give an immense advantage to
> one tribe over another. A tribe including many members
> who, from possessing a high degree of the spirit of patri-
> otism, fidelity, obedience, courage, and sympathy were
> always ready to aid one another, and to sacrifice them-
> selves for the common good, would be victorious over
> most other tribes, and this would be natural selection.
> At all times throughout the world tribes have supplanted

> other tribes; and as morality is one important element in
> their success, the standard of morality and the number
> of well-endowed men will thus everywhere tend to rise
> and increase.[1]

Darwin's reasoning in terms of relative fitness within and among groups is impeccable. He needs to be forgiven for omitting women from his scenario. Also, he doesn't comment on the glaring fact that his scenario only explains the evolution of moral conduct *within* tribes and sets the stage for immoral conduct *toward other tribes*. If we want to explain moral conduct among tribes, in terms of multilevel selection theory, we would need to add another layer to the hierarchy of groups within groups within groups.

The previous chapter showed that Darwin was on the right track, but it has taken a long time for modern evolutionists to reach this conclusion. During the 1960s, a consensus arose that while the first and second foundational principles outlined in the previous chapter are correct, between-group selection is invariably weak compared to within-group selection.[2] If this consensus was correct, then altruism as defined in the last chapter does not exist and behaviors that appear altruistic must be explained another way. Efforts to explain the evolution of altruism "without invoking group selection" went by names such as *inclusive fitness* (also called *kin selection*) *theory*, *selfish gene theory*, and *evolutionary game theory*. These theories had a way of transmuting altruism into selfishness. A relative helping another relative became an individual helping its genes in the body of another individual, thereby maximizing its own "inclusive fitness." Evolutionary game theory rendered altruism as a matter of scratching your back so that you'll scratch mine. Selfish gene theory performed the ultimate transmutation of calling everything that evolves by genetic evolution a form of selfishness. The zeal with which evolutionists during this period declared an end to altruism is captured by these two passages from Michael Ghiselin in 1974 and Richard Alexander in 1987:

> No hint of genuine charity ameliorates our vision of society, once sentimentalism has been laid aside. What passes for co-operation turns out to be a mixture of opportunism and exploitation. . . . Given a full chance to act in his own interest, nothing but expediency will restrain [a person] from brutalizing, from maiming, from murdering—his brother, his mate, his parent, or his child. Scratch an "altruist" and watch a "hypocrite" bleed.

> I suspect that nearly all humans believe it is a normal part of the functioning of every human individual now and then to assist someone else in the realization of that person's own interests to the actual net expense of the altruist. What this "greatest intellectual revolution of the century" tells us is that, despite our intuitions, there is not a shred of evidence to support this view of beneficence, and a great deal of convincing theory suggests that any such view will eventually be judged false. This implies that we will have to start all over again to describe and understand ourselves, in terms alien to our intuitions, and in one way or another different from every discussion of this topic across the whole of human history.[3]

Starting in the mid-1970s, a revelation began to dawn. The theories that claimed to explain altruism without invoking group selection turned out to invoke group selection after all, in every way except using the name.[4] They all assume that social interactions take place in groups that are small compared to the total population. The traits that became labeled only apparently altruistic do not have the highest relative fitness within groups and evolve only by virtue of the differential contribution of the groups to the total population. These theories aren't wrong when it comes to explaining when a given trait evolves in the total population, but they are wrong in denying the role of between-group selection in the evolution of the given trait. The transmutation of altruism into selfishness is therefore based on a difference in how these terms are defined and not a difference in the *causal processes* invoked to explain the evolution of the traits.[5]

The account of natural selection within and among groups that I provided in the last chapter can ultimately be accepted at face value. The controversy over group selection is receding into the past and eventually will be forgotten except from a historical perspective, like the controversies over the Copernican view of the solar system, Darwin's theory of natural selection during the late nineteenth and early twentieth centuries, and the theory of continental drift during the early twentieth century. For this reason I am able to offer a postresolution explanation for the evolution of altruism in a short space—as in the previous chapter, as I promised to do in the Introduction.

Still, just because those other theories failed to provide an alternative causal mechanism for the evolution of altruism doesn't mean they lack merit. Providing a different perspective on the same causal process can be a virtue in its own right. In addition, it is interesting and worth exploring how a behavior that appears altruistic from one perspective can be made to appear selfish from another perspective. This chapter is therefore devoted to the concept of equivalence that has become a lively topic of debate among philosophers of biology, largely (although not exclusively) based on the group selection controversy.[6]

Science is supposed to be a contest of causal explanations. Alternative hypotheses are proposed that invoke different causal processes to explain a given phenomenon. The hypotheses make different predictions about observable phenomena that can be pitted against each other by observation or experiment. The hypothesis that loses the contest is rejected and the winner is provisionally accepted, at least until another hypothesis is proposed. In this fashion, science converges upon the truth—or at least a facsimile that cannot be empirically distinguished from the truth.[7]

This image of science as a relatively straight path to the truth is complicated by the concept of paradigms. Thomas Kuhn[8] observed that scientists sometimes get stuck viewing a topic a certain way. Their particular configuration of ideas is capable of a limited degree of change through hypothesis formation and testing, but cannot escape from their own assumptions in other respects, which makes the replacement of one configuration of ideas (a paradigm) by another configuration a messy and uncertain process.

In the typical rendering of paradigms, one configuration of ideas eventually does replace another, even if the process is long and protracted. Nobody talks about pre-Copernican views of the universe anymore (except from a historical perspective). The concept of equivalence differs from the standard rendering of paradigms in two respects. First, different configurations of ideas exist, as with the standard rendering, *but they deserve to coexist* rather than replacing each other. Second, the different configurations *can be, but need not be, incommensurable.* A single person or an entire community of people can master more than one configuration.

Three analogies are often employed to explain the concept of equivalence: accounting methods, perspectives, and languages. Starting with accounting methods, imagine creating a spreadsheet to organize your finances. In some ways it is useful to organize your expenditures in terms of date. In other ways it is useful to organize them in terms of tax-relevant categories. Both are equally correct and deserve to coexist because they are useful for different purposes. A person accustomed to one accounting method might be confused by another, but can master both with a little effort. With suitable coordination, different accounting methods can be fully intertranslatable.

Proceeding to the analogy of perspectives, imagine that you are planning to ascend a mountain with a friend. As you survey its contours, certain features are difficult to make out from where you are standing. It makes sense for your friend to stand in a different spot so that you can collectively get a better view of the mountain's contours. Recognizing any given feature from both vantage points might be confusing at first, but can become familiar with sufficient effort. The two perspectives are fully intertranslatable because, after all, you are both viewing the same mountain.

Finally, groups of people who become isolated from each other often end up speaking different languages. Different languages are not equivalent at the level of single words; if they were, then translating between languages would be easy. Instead, languages parse the world in different ways. A concept represented by a single word in one language can require a treatise to understand in a different language. If the entire world spoke a single language, then a precious form of diversity would be lost. A person fluent in one language can be mystified by other languages, yet it is also possible to become multilingual; children exposed to different languages can speak and comprehend them spontaneously, while also keeping them separate. Entire populations can be multilingual, which is not just a recent result of admixture but also can be an equilibrium situation. In some cultures, multiple languages are spoken within a small geographical area and people speaking different languages routinely intermarry without the languages blending into a single language.[9]

The examples of accounting methods, perspectives, and languages make it easy to understand how different configurations of thought can be *different but worthy of coexistence*, and how they *can be, but need not be, incommensurate*. The concept of equivalence notes that different configurations of scientific thought, which are variously called *theories, frameworks*, or *paradigms*, can also share these two properties. The concept of equivalence does not replace the standard rendering of paradigms. Some configurations of thought are just plain wrong and deserve to be tossed in the dustbin of history. And the good old-fashioned process of hypothesis formation and testing still takes place within each paradigm. The concept of equivalence needs to be added to these other well-worn concepts. In addition, causally equivalent paradigms need to be carefully distinguished from paradigms that invoke different causal processes. If equivalent paradigms are treated as if one deserves to replace another, then nothing but confusion can result.

Precisely this confusion has plagued evolutionary theories of social behavior for over half a century. Let's return to my example in the last chapter involving A-types that provide a public good and S-types that do nothing to see how different accounting methods can exist for the same evolutionary process. We need some numbers and calculations for precision, but they are no more complicated than balancing your checkbook, so don't be afraid!

Table 3.1 follows the fate of two groups. Both groups start out with ten individuals. Group 1 has 20 percent A-types and Group 2 has 80 percent A-types. Everyone starts out with a baseline fitness of 1. A-types increase the fitness of everyone in their group, including themselves, by 0.2 at a private cost of 0.3. Notice that I have made the cost sufficiently high so that A-types decrease their own absolute fitness (0.2 – 0.3 = –0.1) in the process of benefitting others. If that's not altruism defined by action, what would be?

TABLE 3.1. MULTILEVEL ACCOUNTING METHOD

|  | Group 1 | Group 2 |
|---|---|---|
| Initial group size | 10 | 10 |
| Initial frequency of A | 0.2 | 0.8 |
| Fitness of A | 1.1 | 2.3 |
| Fitness of S | 1.4 | 2.6 |
| Frequency of A after selection | 0.164 | 0.779 |
| Total offspring number | 13.4 | 23.6 |

As in the biological examples of the previous chapter, selfishness beats altruism within each group: S-types have the highest relative fitness, and the frequency of A-types declines from 20 percent to 16.4 percent in Group 1 and from 80 percent to 77.9 percent in Group 2. But altruistic groups beat selfish groups: The group with 80 percent A-types has produced 23.6 offspring, while the group with 20 percent A-types has produced only 13.4 offspring. Between-group selection is stronger than within-group selection, and the frequency of A in the total population increases from 50 percent to 55.7 percent, despite having declined within each group. Altruism is evolving by group selection, despite its selective disadvantage within groups.

I rigged the numbers to produce this result by creating a lot of variation between groups to offset the fairly large fitness difference between A-types and S-types within each group. By fiddling with the numbers, I could get the S-types to increase in frequency in the total population on the strength of their within-group advantage, despite their selective disadvantage between groups. The point of the example is not to show that altruism can evolve—we have already seen that it can in the previous chapter—but to illustrate the logic of multilevel selection as a method of accounting for evolutionary change. Like a financial accountant, I have toted up the fitness costs and benefits in a certain way to arrive at the bottom line of what evolves in the total population.

The perceptive reader might challenge the legitimacy of calculating evolutionary change at the level of the total population. After all, if the groups remain isolated for a sufficient number of generations, A-types will go extinct in both groups. Who cares if one group became larger along the way? The answer is that groups often don't remain isolated for that long. They mix in a variety of ways, as we saw with the water strider and phage virus examples in the previous chapter. The details of the multigroup population structure matter for the net effect of within- vs. between-group selection, but for anything short of total isolation, the differential contribution of groups to the total population must be factored into the equation.

Now let's tally the costs and benefits a different way, by averaging the fitness of individuals across groups, as shown in Table 3.2. Two A-types in Group 1 have a fitness of 1.1 and eight A-types in Group 2 have a fitness of 2.3, for an average fitness of 2.06. Eight S-types

---

**TABLE 3.2.** AVERAGING ACOCUNTING METHOD

---

| | |
|---|---|
| Fitness of average A-type | $2*1.1 + 8*2.3 = 2.06$ |
| Fitness of average S-type | $8*1.4 + 2*2.6 = 1.64$ |
| Original frequency of A | 0.5 |
| Frequency of A after selection | 0.557 |

---

in Group 1 have a fitness of 1.4 and two S-types in Group 2 have a fitness of 2.6, for an average fitness of 1.64. The average fitness of A is greater than the average fitness of S, so A-types increase in frequency in the total population. This method calculates the relative fitness of individuals in the total population without first partitioning relative fitness into within- and between-group components as an intermediate step.

The two methods of accounting for evolutionary change in Tables 3.1 and 3.2 are equivalent in the sense that they make the same biological assumptions and reach the same conclusion about which type evolves in the total population. They differ in the amount of detail that they provide about fitness differentials within and among groups. Averaging the fitness of A- and S-types across groups (the second method) results in a loss of information. We know that the A-types have increased in frequency in the total population, but we don't know whether they evolved on the strength of a within-group or a between-group advantage. The information isn't really lost, of course. If we queried someone using the second method about relative fitness within groups, he or she could retrieve the information. It's just that the information is lost when fitnesses are averaged across groups.

Similar examples of summary statistics that lose information abound. Suppose that you weigh one hundred pigs and report your results in terms of a mean and variance. These summary statistics result in a loss of information. Insofar as different distributions of weight can result in the same mean and variance, you can't re-create a given distribution from its mean and variance. Or suppose that you are calculating the motion of an object that is being subjected to different forces. You can represent each force as a vector and calculate a single resultant vector from the component vectors, but this approach results in a loss of information. If all you know is that the object is moving to the right, you don't know if it is being pushed from the left or pushed from both left and right, but harder from the left.

Against this background, suppose we want to know whether A-types evolve on the strength of a relative fitness advantage within or between groups. The accounting method in Table 3.1 is designed to answer this question and enables us to conclude that A-types evolve on the strength of a between-group advantage, despite their selective disadvantage within groups. There is a right or wrong about the matter, and anyone who claimed that A-types evolve on the strength of a within-group selective advantage would be just plain wrong. This is scientific hypothesis formation and testing in the good old-fashioned sense of the term.

The accounting method in Table 3.2 might be useful for some purposes, but it is not designed to answer the particular question of whether A-types evolve on the strength of a within-group vs. a between-group advantage. Someone who employs this accounting method would have to go back to the original information to make this particular comparison. The situation is comparable to people who organize their finances in terms of dates being asked to calculate their business expenses. The information isn't lost (hopefully!) but needs to be organized a different way.

Now we are in a position to explain the great error that occurred with the rejection of group selection in the 1960s. People used accounting methods similar to Table 3.2, which resulted in a loss of information about relative fitness within and among groups. Then they used the results to claim that their model could explain the evolution of altruism (now modified with the adjective "apparent") without needing to invoke group selection as defined by multilevel selection theory. Elliott Sober and I dubbed this "The Averaging Fallacy" in our 1998 book, *Unto Others*.

Supposing that smart people could be misled in this way might seem far-fetched, but that is because I carefully introduced the concept of equivalence prior to presenting the two examples. Most scientists do not keep the concept of equivalence in mind and tend to assume more uniformity in the definitions of their terms than they should. Here is a brief description of the major theoretical frameworks that were developed to explain the evolution of altruism without invoking group selection, even though fitness differences within and among groups are there for anyone to see, once we know what to look for.

*Evolutionary game theory* averages the fitness (= payoff) of individuals across groups, similar to the accounting method in Table 3.2. For example, in two-person game theory models, individuals interact in pairs. Noncooperative strategies typically beat cooperative strategies within a mixed pair, but pairs of cooperators have a higher combined payoff than mixed pairs or pairs of noncooperators. This is the standard within- vs. between-group selection dynamic, as interpreted by multilevel selection theory, but evolutionary game theory models typically calculate the average payoffs of the strategies across pairs. When the average cooperator is more fit than the average noncooperator, this is interpreted as "individual-level selection" and the winning strategy is called selfish. This argument against group-level selection is no more valid than saying that Table 3.2 is an argument against Table 3.1. As for two-person game theory, so also for N-person game theory.

Evolutionary game theory models typically assume that the phenotypic strategies breed true in an unspecified way.[10] For any evolutionary model that includes genetics, the fitness of genes can be averaged across individuals in the same way that the fitness of individuals can be averaged across groups, which provides the basis for *selfish gene theory*. As an example, consider two genes at a single locus, A and a, giving rise to three genotypes in a diploid species (AA, Aa, and aa). The A gene improves the survival of the whole organism compared to a, so that AA is more fit than Aa, which in turn is more fit than aa. There is no difference in the fitness of A and a genes within heterozygotes. This would be "between-organism selection," according to multilevel selection theory. Selfish gene theory averages the fitness of the A and a genes across organisms in the same way that individual payoffs are averaged across pairs in two-person game theory. A is more fit than a on average, so it evolves by "gene-level selection." The fitness of genes can be averaged across groups in addition to individuals, collapsing anything that evolves at any level (as defined by multilevel selection theory) into gene-level selection (as defined by selfish gene theory). This accounting method might be useful for some purposes, but it was a great error to conclude that "group selection is wrong because the gene is the fundamental unit of selection"— the received wisdom for decades.

*Inclusive fitness theory* (also called *kin selection theory*) calculates the effect of a focal individual's behavior on itself and others, weighted by a coefficient of relatedness. The value of the coefficient is 1 if the recipient is certain to share the same gene causing the behavior in the actor (e.g., an identical twin), and 0 if the recipient is no more likely to share the same gene than an individual drawn at random from the total population. Originally, the coefficient was interpreted in terms of genealogical relatedness (hence the term *kin selection*) but later it was generalized to include genetic correlations for any reason, such as the kind of conditional movements exhibited by water striders described in chapter 2. In any case, inclusive fitness is defined as the sum of the effects on self and others, weighted by the coefficients of relatedness, and the trait evolves when an individual's inclusive fitness is positive.

All of these theories correctly account for what evolves in the total population (given their assumptions) and are intertranslatable in the same way as the examples provided in Tables 3.1 and 3.2. Each can be insightful in the same way as multiple accounting systems, perspectives, and languages outside of science. They therefore deserve to coexist, but for this kind of equivalence to be productive, it is essential for users of the theories to be appropriately multilingual. Otherwise, equivalent paradigms will be pitted pointlessly against each other, as they have for over half a century for the topic of between-group selection.

In my opinion, the concept of equivalence should be part of the basic training of all scientists, along with the concept of paradigms that replace each other and the process of hypothesis formation and testing that takes place within paradigms. Scientists should routinely perform "equivalence checks" to determine whether different paradigms invoke different processes, such that one can be right and the other wrong, or whether they merely reflect different accounting methods (or perspectives, or languages) that invoke the same processes. The amount of time and effort saved avoiding pointless controversy would be colossal.

The need to keep equivalence in mind is especially important for the study of altruism, because each of the major theoretical frameworks divides the pie of social behaviors into seemingly selfish and altruistic slices in different ways. For multilevel selection theory, the division is based on relative fitness within and among groups. For inclusive fitness theory, the division is based on whether an individual incurs an absolute net cost in providing benefits to others. For evolutionary game theory and selfish gene theory, everything that evolves is selfish because it had a higher fitness than what didn't evolve.

To appreciate the emptiness of this last definition, let's go back to the pure altruist whom we imagined at the beginning of chapter 1. Suppose that she is lucky enough to be surrounded by other pure altruists, avoiding the depredations of more selfish people. She therefore thrives, not by virtue of helping herself, but by virtue of being helped by other altruists. Does the fact that she thrives make her selfish? Of course not, yet that is precisely how the so-called alternatives to group selection performed their alchemy of transmuting altruism into selfishness with their definitions.

Now that we have become familiar with how altruism and group-level functional organization evolves, when defined in terms of action and relative fitness differences within and among groups, we are in a position to consider the evolution of altruism and group-level functional organization in our own species.[11]

## CHAPTER 4

# From Nonhumans to Humans

S O FAR, my use of evolution as a navigational guide has been based entirely on genetic evolution. That might suffice for birds, bees, and bacteria, but thinkers throughout history have regarded humans as uniquely different from other species. A common formulation during the post-Darwinian era is that evolution explains the rest of life, our physical bodies, and a few basic urges such as to eat and have sex, but has little to say about our rich behavioral and cultural diversity.

Our quest to answer the question "Does altruism exist?" therefore requires a consideration of humans per se and our distinctive properties in addition to what we share with other species. Fortunately, progress during the last few decades has enabled us to provide an account of human evolution that does justice to our distinctive capacity for behavioral and cultural change, while remaining firmly within the orbit of evolutionary science.

Our starting point is the concept of major evolutionary transitions, which notes that the balance between levels of selection (as defined by multilevel selection theory) is not static but can itself evolve. On rare occasions, mechanisms evolve that suppress disruptive forms of selection within groups, causing benign forms of within-group selection and between-group selection to become the dominant evolutionary forces. When this happens, the groups become so functionally organized that they qualify as organisms in their own right. This concept was originally invoked to explain the origin of nucleated cells and was later generalized to explain the first bacterial cells, multicellular organisms, eusocial insect colonies and possibly even the origin of life itself, as I recounted in chapter 2.

Major evolutionary transitions have three hallmarks. First, they are *rare events* in the history of life. Group-level selection figures in the evolution of many single traits in many species, as we have seen from the examples in chapter 2, but becoming the primary evolutionary force for most traits in a species is another matter.

Second, major evolutionary transitions have *momentous consequences* once they occur. Higher-level superorganisms such as nucleated cells, multicellular organisms, and eusocial insect colonies dominate their lower-level competitors (bacterial cells, single nucleated cells, and solitary insect species, respectively) in ecological competition. According to current estimates, eusocial insect societies (primarily wasps, bees, ants, and termites) originated only about a dozen times but currently account for over half of the insect biomass on earth.[1] The lower-level units of functional organization are not completely displaced. There are still plenty of bacteria, single-celled nucleated organisms, and solitary insect species on the face of the earth, illustrating the important point that higher-level functional organization is not adaptive under all circumstances. Nevertheless, when an ant colony moves into a rotten log, most of the solitary invertebrate species in the vicinity are quickly displaced.

Third, the suppression of disruptive forms of within-group selection is only partial and not complete. Pure organisms, whose lower-level elements work 100 percent for the common good, do not exist. One of the most important discoveries in evolutionary biology during the last few decades has been to realize how much a multicellular organism is a highly regulated society of cells that is elaborately organized to withstand an onslaught of cheating and exploitation from within.[2]

Against this background, our distinctiveness as a species can be summarized in a single sentence: *We are evolution's latest major transition.* Alone among primate species, we crossed the threshold from groups *of* organisms to groups *as* organisms. Other primate species cooperate to a degree, sometimes to an impressive degree, but disruptive within-group competition for mates and resources is still a strong evolutionary force.[3] Even the cooperation that does take place within primate groups often consists of coalitions competing against other coalitions within the same group. Our ancestors managed to suppress disruptive forms of within-group competition, making benign forms of within-group selection and between-group selection the primary evolutionary forces.[4]

The distinction between disruptive and benign forms of within-group selection is crucial. There's a world of difference between socially dominant individuals in most primate groups, who simply appropriate the best mates and resources for themselves, and high-status individuals in small-scale human societies, who must earn their status by cultivating a good reputation. The kind of social control that suppresses destructive within-group competition but permits and often cultivates group-beneficial forms of within-group competition is part of what the concept of major evolutionary transitions is all about.[5]

All of the hallmarks of a major evolutionary transition are present in the human case. It was a rare event, happening only once among primates, and the combination of a species that is both functionally organized at the group level and highly intelligent at the individual level is doubly rare, as we shall see. It had momentous consequences. Just as eusocial insects constitute over half of the insect biomass on earth, we and our domesticated animals represent a large fraction of the vertebrate biomass on earth, for better or for worse.[6] The suppression of disruptive forms of lower-level selection is only partial and by no means complete. Everyday life and the annals of history are replete with examples of individuals and factions that succeed at the expense of their groups, despite the arsenal of social control mechanisms designed to thwart them. If the cells of multicellular organisms could talk, they would tell stories similar to the ones that we tell each other.

The idea that a human society is like a single body or a social insect colony is both old and new. Religious believers around the world are fond of comparing their communities to bodies and beehives, as I recount in chapter 6 and at greater length in my book *Darwin's Cathedral*.[7] The founders of the human social sciences also took the organismic concept of human society seriously. As Harvard psychologist Daniel M. Wegner puts it,

> Social commentators once found it very useful to analyze the behavior of groups by the same expedient used in analyzing the behavior of individuals. The group, like the person, was assumed to be sentient, to have a form of mental activity that guides action. Rousseau and Hegel were the early architects of this form of analysis, and it became so widely used in the 19th and early 20th centuries that almost every early social theorist we now recognize as a contributor to modern social psychology held a similar view.[8]

Yet this view was largely eclipsed by a paradigm shift that took place during the middle of the twentieth century that is often labeled *methodological individualism*. According to social psychologist Donald Campbell,

> Methodological individualism dominates our neighboring field of economics, much of sociology, and all of psychology's excursions into organizational theory. This is the dogma that all human social group processes are to be explained by laws of individual behavior—that groups and social organizations have no ontological reality—that where used, references to organizations, etc. are but convenient summaries of individual behavior.[9]

This kind of individualism was not restricted to the ivory tower but came to permeate modern everyday life. As British prime minister Margaret Thatcher famously commented during an interview in 1987, "There is no such thing as society. There are individual men and women, and there are families." I have more to say about individualism in political and economic thought in chapter 7.

Against the background of individualism, the prospect that human societies are like bodies and beehives after all is a paradigm shift of the first rank. Fortunately, we have a rich early tradition in the human social sciences to consult in addition to a more recent literature informed by evolutionary theory.

The functional organization of human groups during our evolutionary past included physical activities such as childcare, food acquisition, predator defense, and trade and warfare with other groups. It also included mental activities, which is one reason that I featured the concept of mentality as a group-level adaptation in chapter 1. In fact, most of the mental attributes that we regard as distinctively human, such as our capacity for symbolic thought (including but not restricted to language) and the ability to transmit learned information across generations (culture), are fundamentally communal activities. This assertion has led to a hypothesis that a *single* shift in the balance between levels of selection led to the *entire package* of distinctively human traits, including our ability to cooperate in groups of unrelated individuals, our distinctive cognition, and our ability to transmit culture. I have called this the *cooperation came first* hypothesis,[10] although I now prefer the phrase *group-level functional organization came first*, to underscore the fact that group-level functional organization need not look like overt cooperation, any more than it needs to look like altruism.

According to this hypothesis, most primate species are very smart as individuals, but their intelligence is predicated upon distrust. Chimpanzees, our closest living relatives, rival and even exceed our intelligence in some respects but have mental deficits that seem strange to us, such as the ability to understand the information conveyed in pointing.[11] In some respects, dogs have more humanlike intelligence than apes, because dogs have been genetically coevolving with humans for thousands of generations and apes have not.[12] This fact seems amazing to those who think of braininess as a single trait, with humans the brainiest species, our closest ape relatives distant seconds, and dogs somewhere in the middle of the pack. It becomes more sensible when we regard the brains of all species as adapted to their respective environments and our brains as adapted to life in groups whose members could be trusted, for the most part, to act on our behalf. The brains of any species evolve in certain ways in trustworthy social environments and other ways in treacherous environments—hence the outcome that a distantly related species such as the dog can resemble our intelligence more than a closely related species such as the chimpanzee.

Our capacity for symbolic thought that can be transmitted across generations, including but not restricted to language, led to a quantum jump in our ability to adapt to our current environments. Many species form mental relations that correspond closely to environmental relations, such as associating food with a sound immediately preceding the presentation of the food. However, these mental relations are broken as easily as they are formed. Symbolic thought involves the creation of mental relations that persist in the absence of corresponding environmental relations.[13] Rats associate the word "cheese" with the food as long as the two are presented together, but not otherwise. In contrast, I could say the word "cheese" to you a million times without presenting cheese, and the mental relation would still persist. Humans even have symbols for imaginary entities, such as "trolls," that don't exist in the real world.

The ability to create symbolic relations that don't correspond to environmental relations might seem maladaptive, but symbolic relations remain connected to the environment in another sense. Every suite of symbolic relations motivates a suite of actions that can potentially influence survival and reproduction in the real world. If we call a given suite of symbolic relations a "symbotype," then there is a symbotype-phenotype relationship comparable to a genotype-phenotype relationship.[14] Moreover, just as genetic polymorphisms at many loci result in a nearly infinite number of genotypic combinations, permutations of symbolic relations result in a nearly infinite number of symbotypic combinations, each with potentially a different effect upon human action. In short, the human capacity for symbolic thought is nothing less than a new system of inheritance.[15]

Although the concept of symbolic thought as an inheritance system is simple enough to grasp, it is a seismic shift in thinking against various intellectual backgrounds. Darwin knew nothing about genes and conceptualized inheritance as any mechanism that creates a resemblance between parents and offspring. Once genes were discovered, however, they became the only mechanism of inheritance in the minds of most evolutionary biologists and the general public. Inside and outside the ivory tower, say the word "evolution" and most people hear the word "genes."[16] Current evolutionary theory therefore needs to expand to include the concept of symbolic thought as a nongenetic inheritance system.

A different seismic shift takes place for intellectual traditions that are centered on symbolic thought, but not from an evolutionary perspective. Thinkers from traditions such as social constructivism and postmodernism also tend to associate evolution exclusively with genetic evolution, which causes them to regard their own constructs as outside the orbit of evolution altogether. For them, the need to study symbolic thought from a modern evolutionary perspective might take some adjustment,[17] yet they have as much to teach evolutionary biologists as to learn from them.

Much is sometimes made of the fact that human behavioral and cultural processes are often directed toward a goal, in contrast to genetic variation, which is not random in the strict sense of the word but is said to be arbitrary with respect to the traits that are selected. The terms *Lamarckian* and *Darwinian* are often used to make this comparison, but avoiding historical revisionism is important. Darwin also invoked the inheritance of acquired characters in his effort to explain the nature of heritable variation. Moreover, as Eva Jablonka and Marion Lamb explain in their important book *Evolution in Four Dimensions*, if Lamarck had been correct, the outcome of evolution would be much the same—there would still be giraffes with long necks, for example. In fact, at least some forms of genetic variation are proving to be directed after all.[18]

Theoretically, there is nothing heretical or "non-Darwinian" about goal-directed evolutionary processes, because their directed aspects evolved by undirected heritable variation in the past. Moreover, goal-directed processes typically include undirected components. Consider intentional decisionmaking, one of the most goal-directed forms of behavioral and cultural change, which involves explicitly selecting among alternative options with set criteria in mind. The search for options can be either directed or undirected, and the most creative options often come out of nowhere. That's what brainstorming and thinking out of the box are all about. Evolutionary algorithms have become important engineering tools because having a computer randomly generate options can identify potential solutions better than more narrow goal-directed algorithms. In short, if we regard decisionmaking as an explicit variation-and-selection process, the variation part often benefits from an undirected component.

The selection part of an intentional decisionmaking process is goal-directed by definition, but many selection criteria are possible. Some individuals might select the options that maximize their relative advantage within their groups. Others might select the options that maximize the advantage of their groups, compared to other groups. Others might select the options that maximize world peace or the sustainability of the planet. Which selection criteria come to be employed in any particular decisionmaking process? Another selection process must be invoked to answer this question. In this fashion, current variation-and-selection processes must be explained in terms of past variation-and-selection processes, like peeling away the layers of an onion. In addition to conscious decisionmaking, other directed selection processes take place subliminally, such as our tendency to copy the behaviors of high-status individuals.[19] Then there is the raw process of undirected cultural evolution—many inadvertent social experiments, a few that succeed. Even intentional decisions result in variation that is arbitrary with respect to what is selected when they produce unintended consequences or collide with each other.

The most important common denominator for variation-and-selection processes, whether directed or undirected, is that they are *open-ended*. They are capable of producing new phenotypes in response to current environmental conditions. This is in contrast to what evolutionists call *closed phenotypic plasticity*, which selects among a fixed repertoire of phenotypes in response to current environmental conditions.[20] The best way to appreciate the open nature of human phenotypic plasticity is to step back from the "trees" of the academic literature to view the "forest" of the human social conquest of earth, as E. O. Wilson puts it.[21] A single biological species spread out of Africa and inhabited the globe, adapting to all climatic zones and occupying hundreds of ecological niches, in just tens of thousands of years. Each culture has mental and physical toolkits for survival and reproduction that no individual could possibly learn in a lifetime. Then the advent of agriculture enabled the scale of human society to increase by many orders of magnitude, resulting in megasocieties unlike anything our species had previously experienced. The human cultural adaptive radiation is comparable in scope to the genetic adaptive radiations of major taxonomic groups such as mammals and dinosaurs.[22] What else is required to conclude that symbolic thought functions as an inheritance system comparable to the genetic inheritance system?

Despite important differences between the two inheritance systems, most of the core principles that I have described for genetic evolution in previous chapters apply to both, especially with respect to multilevel selection. Regardless of whether a phenotypic trait is genetically inherited, learned, or culturally derived, it can spread by virtue of benefitting individuals compared to other individuals in the same group, by benefitting all individuals in a group compared to other groups, and so on for a multilevel hierarchy of groups. Extending the hierarchy downward, cultural traits can spread at the expense of individuals, similar to cancer cells and meiotic drive genes.[23] The basic tradeoffs that create conflicts between levels of selection do not depend upon the mode of inheritance or the distinction between directed vs. undirected variation.

Against this background, human history can be seen as a fossil record of multilevel cultural evolution with major transitions of its own. Thousands of generations of gene-culture coevolution equipped our ancestors with a sophisticated ability to function as corporate units at relatively small social and spatial scales. The genetic architecture of our minds was shaped by this process. Then the invention of agriculture made larger groups possible, but the mechanisms that enable small groups to function as corporate units do not necessarily scale up.[24] New culturally derived mechanisms were required that interfaced with older genetic and cultural mechanisms.[25] Evolution is inherently a path-dependent historical process, so the culturally derived mechanisms of large-scale functional organization that evolved in some regions of the world need not resemble those that evolved in other regions of the world. For example, the nation-states of Europe are the product of centuries of between-group military and economic competition, which might or might not serve as viable models for Africa or the Middle East.[26]

Now at last we can proceed from the study of altruism defined in terms of action to the study of altruism defined in terms of thoughts and feelings.

## CHAPTER 5

# Psychological Altruism

IN THE 1980s the philosopher Elliott Sober contacted me to discuss the seemingly nonphilosophical topic of sex ratios. After years of basing philosophy of science on physics, philosophers had decided that biology also warranted their attention. Elliott was writing a book entitled *The Nature of Selection*[1] and discovered on his own that evolutionists had taken a wrong turn in their rejection of group selection. Exhibit A was the proportion of sons and daughters that sexually reproducing organisms produce, which in principle can vary from all sons to all daughters. The details need not concern us here, but sex ratio was regarded at the time as the strongest evidence that between-group selection is invariably weak compared to within-group selection. George C. Williams, whose book *Adaptation and Natural Selection*[2] seemed to sound the death knell for group selection, used sex ratio as his showcase example and concluded his book by writing, "I regard the problem of sex ratio as solved." Elliott begged to differ and was excited to discover that Robert K. Colwell and I were also challenging Williams's interpretation.[3] (Williams eventually reversed his position on sex ratio and other traits, such as the evolution of avirulence in disease organisms.[4])

I was thrilled to talk to a philosopher about the big questions posed by evolutionary theory. Everyone likes to philosophize (at least I do!), but who has the opportunity to bounce ideas around with a world-class philosopher? Elliott must have felt the same way about having access to an evolutionist, because we have been collaborating and keeping in touch ever since. Two major articles that we authored together were titled "Reviving the Superorganism" in 1989 and "Reintroducing Group Selection to the Human Behavioral Sciences" in 1994. Then Elliott decided to write a book on altruism and asked me to join him, resulting in *Unto Others: The Evolution and Psychology of Unselfish Behavior* in 1998.[5] As these titles imply, we were helping to lay the groundwork for the new consensus about multilevel selection that is reported in this book.

*Unto Others* is divided into a section on altruism at the level of action and a section on altruism at the level of thoughts and feelings. I had primary responsibility for the first section and Elliott had primary responsibility for the second, although we also pored over each other's sections. Unsurprisingly, Elliott based his section on the long history of philosophical thought on altruism, dating back to figures such as John Stuart Mill and Jeremy Bentham, who I learned had himself stuffed so that his corpse could preside over meetings after his death. Elliott also covered the modern psychological literature on altruism, including work by Daniel Batson, who conducted a series of clever experiments to distinguish subtle forms of psychological egoism from more genuine psychological altruism.[6]

I learned a lot and admired Elliott's handling of the issues, but I couldn't help wondering if the first thinkers had taken a wrong turn and that, by following in their footsteps, more recent inquiry was on the wrong path. Sticking to distinctions such as hedonism, egoism, and altruism seemed antiquated, since these concepts predated not only evolutionary theory but also the emergence of psychology as a science. Much of the recent philosophical and psychological literature seemed like a parlor game in which proponents of selfishness described a hypothetical psychological mechanism that did not count as altruistic according to their criteria, but which produced apparently altruistic behavior, such as a soldier falling upon a grenade to save his comrades. Proponents of altruism were then required to disprove the hypothetical selfish mechanism—not an easy task, since the distinction often centered on mental events that are difficult to observe. A principle of parsimony was often invoked to the effect that selfish psychological mechanisms are simpler than altruistic mechanisms and should be preferred in the absence of other deciding evidence. After a few rounds of playing this game, the selfish psychological mechanisms being discussed were virtually identical to altruistic psychological mechanisms in their behavioral manifestations, which was why cleverness was required on the part of philosophers to tease them apart in thought experiments or of psychologists, such as Batson, to tease them apart in laboratory experiments. The more cleverness was required, the less I cared about the outcome any more than I care whether someone pays me by cash or check. It seemed that the entire enterprise had lost sight of the fact that thoughts and feelings ultimately need to be judged in terms of what they cause people to do!

Shortly after he seized upon the idea of natural selection, Darwin jotted in his private notebook: "he who understands baboon would do more toward metaphysics than Locke." In this chapter, I attempt to show how the key distinction between proximate and ultimate causation in evolutionary theory can help to reorganize the study of thoughts and feelings relevant to the expression of altruism and group-level functional organization at the level of action. I begin with a basic tutorial for readers who are not already familiar with the distinction.

One of the most remarkable features of evolutionary theory is its ability to predict the observable properties of organisms and societies without any knowledge of proximate mechanisms whatsoever. For example, we can confidently predict that many desert species will be sandy colored to avoid detection by their predators and prey. We can make this prediction for snails, insects, reptiles, birds, and mammals, even though they have different genes and physical exteriors. If the desert has black (or white) sand, the species will be black (or white). Predicting observable properties purely on the basis of what enhances survival and reproduction in a given environment is one of the most powerful tools in the evolutionary toolkit.

The reason that these predictions are often (but not always) correct is because of heritable variation. Whenever an observable trait is heritable (such as coloration), then variants that enhance survival and reproduction (such as colors that match the background) increase in frequency in the population until they become species-typical. The material makeup of the organism is important only insofar as it results in heritable variation. As we have seen, very different physical systems can meet this criterion, including the genetic inheritance system, the physical underpinnings of the human symbolic inheritance system, and the physical underpinnings of evolutionary algorithms in computers.

The environmental forces that act upon heritable variation are called *ultimate causation* in evolutionary theory. These forces include natural selection—the root cause of all functional design—and random processes that cause some traits to evolve rather than others based on the luck of the draw. Ultimate causation is contrasted with *proximate causation*, which is needed to explain the physical basis of any given trait. For example, coloration in any given desert insect species is coded by certain genes, which act upon certain physical materials in the developing organisms such as chitin. Coloration in a given desert mammal species is coded by other genes acting upon other physical materials such as keratin. Insofar as any phenotypic trait that evolves has a physical basis, full understanding requires an explanation in terms of both ultimate and proximate causation, which complement and never substitute for each other.

Ernst Mayr[7] is credited with emphasizing the ultimate-proximate distinction, although it has always been implicit in evolutionary theory. Nobel laureate Niko Tinbergen[8] independently called attention to four questions that must be addressed to fully understand any given trait, concerning its function, mechanism, development, and phylogeny. Tinbergen's fourfold distinction adds a temporal dimension to Mayr's twofold distinction. As an example of development as a separate question, the vertebrate eye can be studied as a mechanism of vision in its developed form, without necessarily considering how the eye developed during the lifetime of the organism. The study of development per se is required for fuller understanding. As an example of phylogeny as a separate question, consider the placenta of mammals and the pouch of marsupials. Both have the same function of gestating offspring, but they are very different from each other because mammals and marsupials evolved in isolation from each other on different continents. In general, evolution is a historical process that seldom takes the same path twice, a fact as important for cultural as for genetic evolution.

A number of implications follow from these distinctions that are highly relevant to the study of altruism. The most important implication is that proximate and ultimate causation stand in a many-to-one relationship. Just as there are many ways to skin the proverbial cat, many proximate mechanisms can result in the same functional trait. Some of my previous examples already illustrate this point, such as the cash or check distinction, the many desert species that become sandy colored, and placental and marsupial mammals evolving different solutions to the same functional problem of gestation. Two additional human-related examples nail down this crucial point.

Consider the ability of human adults to digest lactose. For all mammal species except humans during the last few thousand years, milk is a food for infants but not adults. Natural selection has resulted in a digestive physiology that turns off the ability to digest lactose shortly after weaning. When humans started to domesticate livestock some 10,000 years ago and to drink their milk, they were ingesting something to which they were not genetically adapted. Cultural practices such as fermentation (in which microorganisms do the digesting) enabled milk to be utilized without genetic evolution, but then genetic evolution followed suit over the longer term. In other words, mutations that enabled adults to digest lactose arose and spread in cultures that were using milk as an adult food. This is one of the best-documented examples of gene-culture coevolution, in which cultures shape the selection of genes as much as the reverse.

The most important part of the story for our purpose is that dairying practices arose independently in different geographical regions of the world. Although the selection pressure to digest lactose as adults was the same, the particular mutations that arose were not. Insofar as the functional ability to digest lactose can be accomplished by more than one physiological mechanism, the same mechanism would not likely arise in geographically isolated locations.[9] Hence, not only do some human adults lack the ability to digest lactose at all (because their ancestors did not come from dairying cultures), but those who can digest lactose do so in different ways (because their ancestors came from different dairying cultures). Proximate and ultimate causation exist in a many-to-one relationship.

My second example involves groups that attempt to manage common-pool resources such as irrigation systems, pastures, forests, and fisheries. As I briefly described in chapter 1, Elinor Ostrom received the Nobel Prize in economics in 2009 for showing that common-pool resource (CPR) groups are capable of avoiding the tragedy of overuse, but only if they possess certain design principles. This example is directly relevant to the topic of altruism because the prudent use of resources definitely qualifies as altruism at the level of action. The prudent viral strain described in chapter 3 is altruistic compared to the prolific strain. Managing resources for long-term sustainability would be high on the list of the pure altruist, who we imagined in chapter 1, as part of what people must do to make the world a better place. If actual groups have accomplished this goal, then we want to know how they do it, regardless of whether their thoughts and feelings appear altruistic or selfish. If proximate psychological mechanisms that count as selfish can achieve altruism at the level of action better than proximate psychological mechanisms that count as altruistic, then bring on the selfishness!

Here are the eight design principles that I described in chapter 1.

1. Strong group identity and understanding of purpose.
2. Proportional equivalence between benefits and costs.
3. Collective-choice arrangements.
4. Monitoring.
5. Graduated sanctions.

6. Conflict resolution mechanisms.

7. Minimal recognition of rights to organize.

8. For groups that are part of larger social systems, appropriate coordination among relevant groups.

Let's review them from the perspective of three imaginary people, Tom, Dick, and Harry, who have different psychological motives.[10] Tom is a *relative fitness maximizer*. If the common-pool resource is a fishery, he wants more fish than others in his group, regardless of the number of fish. Dick is an *absolute individual fitness maximizer*. He wants as many fish as possible for himself, without caring whether others have more or less than he does. Harry is a *group fitness maximizer*. He wants to maximize the fish harvest for the group, regardless of his proportion of the catch. In conventional terms, Tom and Dick are variants of egoism and Harry is an altruist. How would they regard the eight design principles?

For Tom, most of the design principles would be rules made to be broken. He cares nothing about the purpose of long-term sustainability (1). Making benefits proportionate to costs (2) thwarts his basic purpose, as do the other principles that are designed to promote fairness within the group (3–6). Tom is an outlaw as far as the design principles are concerned.

Dick would find the design principles acceptable to the degree that a rising tide lifts his boat. He's fine with long-term sustainability as long as he gets his share. He might even play a role inventing and implementing the design principles if he thought that the alternative would be ruinous to his catch over the long term. But he would also be tempted to subvert the design principles whenever he could increase his absolute catch at the expense of other members of his group. The design principles are needed to keep people like Dick in solid-citizen mode.[11]

Harry would agree completely with the goal of long-term sustainability (1) and would never be tempted to overexploit the resource for his own gain. Design principles 2–6 wouldn't be necessary if everyone was like Harry.

Now let's compare Tom, Dick, and Harry in a different way, as competitors in a Darwinian struggle for existence. Their fitness would depend critically on the presence or absence of the design principles. Tom would thrive in groups that lack the design principles because of all the opportunities to exploit others for his own gain. He would not thrive in groups that possess the design principles. His kind would either go extinct entirely or be maintained at a low frequency in the population.

Dick can potentially fare well in the presence of the design principles, as long as he operates in solid-citizen mode. He will also fare well in their absence, as long as he operates in exploitation mode. If Dick and Tom were stranded together on a single desert island, Tom would probably win because of his obsession with relative fitness, but in a multigroup environment lacking the design principles, Dick would probably beat Tom.[12]

Harry doesn't stand a chance in groups that lack the design principles, but how does he fare in competition with Dick in groups that possess the design principles? As long as Dick remains in solid-citizen mode, there is *no difference* in how they behave and therefore *no difference in their fitness*. Natural selection would be indifferent as to which type evolves. The fact that Dick is a psychological egoist and Harry is a psychological altruist is irrelevant as far as their evolution is concerned.

Now suppose that Dick, being motivated entirely by self-interest, is occasionally tempted to cheat on the rules imposed by the design principles. This can weigh in his favor if he succeeds or against him if he gets caught. If the design principles are strongly in place, then Dick might become less fit than Harry, who is never tempted to cheat. Harry wins the Darwinian contest—but only if the design principles are strongly in place.

Finally, suppose that we zoom in on Harry and use the latest advances of neurobiology to understand his psychological motives in minute detail. We discover that Harry comes in many varieties. They are all alike at the behavioral level by acting only for the common good, but they differ in their specific motives, like so many ways of skinning the same cat. Unlike the comparison between Dick and Harry, who are behaviorally equivalent under some circumstances but not others, the varieties of Harry are behaviorally equivalent under all circumstances.

So much for thought experiments. What about real CPR groups? One of Elinor Ostrom's most important discoveries was that *CPR groups implement the design principles in different ways that draw upon different motives, norms, and social conventions.* Groups tinker with their arrangements to come up with solutions that work for them. There is no reason to expect all CPR groups to converge upon the same configuration of proximate mechanisms, any more than we should expect the same genetic mutations enabling adults to digest milk to arise in different human populations that keep livestock. The one-to-many relationship between ultimate and proximate causation applies to cultural evolution as much as genetic evolution.

Against this background, we can make some general points about how the proximate mechanisms that lead to altruistic action need to be studied from an evolutionary perspective and how this differs from other perspectives. First, *proximate and ultimate causation need to be studied in conjunction with each other.* This is in contrast to many other perspectives, which focus on proximate causation to the neglect of ultimate causation. All psychological mechanisms that result in altruistic actions exist by virtue of a history of selection or drift. They need to be studied in terms of their functional consequences, their phylogenetic history, and their development during the lifetime of the individual. Very few studies of psychological altruism employ Mayr's twofold or Tinbergen's fourfold approach.

Second, *proximate mechanisms need not resemble functional consequences in any way whatsoever.* They merely need to cause the functional consequences. The functional explanation for why apple trees bloom in spring is that those that bloomed earlier were nipped by frost and those that bloomed later had insufficient time to develop their fruits. The mechanistic explanation for why apple trees bloom in spring is because they are sensitive to day length. Day length per se has no effect on plant fitness; it is just a reliable cue for causing plants to bloom at the right time of year. By the same token, the proximate mechanisms that cause people to behave altruistically, defined in terms of action, need not be motivated by altruism, defined in terms of psychological motives. In principle, successful CPR groups can be composed entirely of people like Dick, not people like Harry. The actual proximate psychological mechanisms that evolve to motivate altruistic actions must be discovered by empirical inquiry guided by Tinbergen's four-question approach.

Third, *cultural evolution plays a strong role in the evolution of proximate mechanisms that motivate altruistic actions, not just genetic evolution.* There is a tendency to assume that a proximate psychological mechanism that operates strongly in one culture must operate in all cultures. This assumption is almost certainly false, given the one-to-many relationship between ultimate and proximate causation. Someone who understands the proximate mechanisms leading to altruistic action in one culture might be like a biologist who understands the genetic and physiological mechanisms leading to cryptic coloration in a particular desert insect species. This knowledge is essential for understanding that species of insect, but becomes limited and even useless for understanding other cryptically colored desert species. Human cultures, like biological species, are historical entities adapted to different environments. If modern CPR groups differ in the proximate mechanisms that result in prudent resource use, what kind of diversity can we expect among major religious traditions or secular forms of governance in the mechanisms that result in group-level functional organization?

Fourth, *the evolutionary fate of a given psychological mechanism that leads to altruistic action depends critically on the environment, including the human-constructed environment.* The core design principles that Elinor Ostrom identified do not seem very altruistic in conventional psychological terms. She could have discovered that the most successful CPR groups are those that have the strongest sense of empathy or have inculcated norms of giving without expectation of gain for oneself. Instead, she discovered that the most successful CPR groups are those that can defend themselves against actions that benefit some individuals at the expense of others within the group. That's what most of the design principles are all about. Without such protections, psychological altruism and any other proximate mechanism that leads to altruistic action are likely to lose the Darwinian contest. Given such protections, altruistic actions can win the Darwinian contest and the question of whether they are motivated by psychological mechanisms that count as altruistic must be decided by empirical inquiry.

To fully understand the proximate mechanisms leading to altruistic action, we need to study the construction of entire social environments, not just what motivates individual people. Jonathan Haidt, an evolutionary social psychologist who is embarking on new studies of the world of business, puts it well when he says that ethics is not something that can be taught to an individual. Ethics is a property of a whole system, a concept that Haidt is trying to substantiate with a website called Ethicalsystems.org.[13]

Let's once again review the progress that we have made in this short book on the big question of altruism. We began by making the crucial distinction between altruism defined in terms of action and altruism defined in terms of thoughts and feelings. We also made a connection between altruism defined in terms of action and group-level functional organization. Improving the welfare of others (the goal of altruism) requires working together to achieve common goals (group-level functional organization). Moreover, we established that functionally organized groups *do* exist, in both human society and the animal world, a fact that can be established with minimal reference to evolution.

In chapter 2 we introduced evolution as an essential theory for explaining why group-level functional organization exists under some circumstances but not others. The reason is due to selection pressures that operate in opposing directions within and between groups. Traits that cause groups to function well are typically different from the traits that maximize relative fitness within groups and require a process of between-group selection to evolve. Just as Rabbi Hillel could recite the meaning of the Torah while standing on one foot, a one-foot version of multilevel selection theory is: "Selfishness beats altruism within groups. Altruistic groups beat selfish groups. Everything else is commentary."

A crucial insight of evolutionary theory in general and multilevel selection theory in particular is that natural selection is based on relative fitness. This is in contrast to many other perspectives that use absolute fitness (or welfare, or utility) as the frame of reference. Defining altruism based on action in terms of relative fitness within and among groups reveals that high-cost forms of group-level functional organization (such as individuals sacrificing their lives for each other) and low-cost forms (such as coordinating activities to make a good collective decision) are different in degree but not kind. In both cases, the traits required for groups to function well typically do not maximize relative fitness within groups and require a process of between-group selection to evolve.

Multilevel selection theory is so simple and well tailored to explain the evolution of group-level functional organization that it is difficult for a newcomer to comprehend its long and controversial history. This places it in the same category as other scientific advances, such as the Copernican view of the solar system, Darwin's theory of evolution, and the theory of continental drift, which are obvious in retrospect but required decades for their resolution. Chapter 3 shows that much of the confusion in the past was based on a failure to distinguish between paradigms that invoke different causal processes, such that one can be right and another wrong, from paradigms that are different but deserve to coexist, like different accounting methods, perspectives, and languages. Once the concept of equivalence is understood, then all evolutionary theories of social behavior can be seen to make the same assumptions about selection within and among groups made by multilevel selection theory. There is no substantive argument against group selection, enabling me to provide a postresolution account in this book.

No account of human altruism is complete without explaining the *differences* between humans and other species in addition to their similarities. We are distinctive in our ability to cooperate in groups of unrelated individuals, in our distinctive forms of cognition that includes symbolic thought, and in our ability to transmit learned information across generations. All of these can be understood as forms of physical and mental teamwork that resulted from a major evolutionary transition, which suppressed disruptive forms of within-group selection and made between-group selection the primary evolutionary force. Teamwork is the signature adaptation of our species. The human major transition justifies a comparison between human social groups, eusocial insect colonies, multicellular and single-celled organisms, and even the origin of life as groups of cooperating molecular interactions. The very concepts of "organism" and "society" have merged, as I showed in chapter 4.

Finally, I began in chapter 5 to tackle altruism defined at the level of thoughts and feelings. I showed that the distinction between proximate and ultimate causation in evolutionary theory leads to a different approach to studying psychological altruism than other perspectives. It is different from the systematizing efforts of philosophers such as Kant, Bentham, and Mills. It is different from the efforts of modern philosophers and psychologists to tease apart subtle differences in thoughts and feelings that have largely the same behavioral manifestations. And it is highly relevant to the altruistic objective of improving human welfare in the real world.

The rest of this book uses the navigational tools that I have provided in the first five chapters to explore altruism in religion, economics, and everyday life. In each case, I first ask whether altruism exists at the level of *action*. In other words, how well do religions, economies, and everyday social units, such as city neighborhoods, function to improve the welfare of their members? Then I explore the proximate mechanisms that cause these groups to function well, to the extent that they do. Do high-functioning groups require people like Harry, who are directly motivated to benefit the common good, or can they draw upon the motives of people like Dick, who cares only about his own absolute welfare? We know from the one-to-many relationship between ultimate and proximate causation that there might not be a single answer to this question. We must examine the empirical evidence on a case-by-case basis and let the chips fall where they may.

CHAPTER 6

# *Altruism and Religion*

I N 1998, one year after the publication of *Unto Others*, the John
Templeton Foundation announced a funding program on the
science of forgiveness. John Templeton admired both religion
and science and thought that they could be mutually supportive.
What could science say about a religious virtue such as "Turn the
other cheek"?

Forgiveness is a religious virtue but it is by no means restricted
to religion. The capacity to forgive under some circumstances
but not others is part of the human psychological repertoire. You
might think that Templeton was a latecomer to the scientific
study of forgiveness, but this was far from the case. Forgiveness
was a backwater subject in psychology, so Templeton was advanc-
ing a frontier of science with his funding initiative, apart from any
religious considerations.

I teamed up with anthropologist Christopher Boehm to write
a proposal titled "Forgiveness from an Evolutionary Perspective."
We would study the human propensity to forgive using evolutionary theory as our navigational guide, much as I am doing for
altruism in this book. Then we would apply the theory to two

large bodies of information. Chris would study forgiveness in hunter-gatherer societies and I would study forgiveness in religious societies around the world. Our proposal was funded and we embarked upon an intellectual adventure that would not have occurred otherwise.

Chris and I had broader objectives in mind than forgiveness. He wanted to organize the entire ethnographic literature on hunter-gatherer societies, which could be used to study many topics in addition to forgiveness. This effort contributed to his superb book *Moral Origins: The Evolution of Virtue, Altruism, and Shame*.[1] I wanted to study religion in general terms, including but not restricted to forgiveness. I had already dabbled in the literature and knew that religious believers are fond of comparing their communities to bodies and beehives, as in this passage from a seventeenth-century Hutterite text (the Hutterites are a Christian sect that originated during the Protestant Reformation and migrated to North America, where it is still thriving):

> True love means growth for the whole organism, whose members are all interdependent and serve each other. That is the outward form of the inner working of the Spirit, the organism of the Body governed by Christ. We see the same thing among the bees, who all work with equal zeal gathering honey.[2]

The concept of major evolutionary transitions meant that this metaphorical comparison could be placed on a solid scientific foundation. If I wanted to study human groups as superorganisms, shouldn't I be studying religious groups? My broad effort resulted in *Darwin's Cathedral: Evolution, Religion, and the Nature of Society*,[3] which was published in 2002 and includes a chapter titled "Forgiveness as a Complex Adaptation," fulfilling the narrow objective of our Templeton Foundation grant.

I was not the only evolutionist to take an interest in religion at the dawn of the twenty-first century. Two other books that appeared within the same year were Pascal Boyer's *Religion Explained* and Scott Atran's *In Gods We Trust: The Evolutionary Landscape of Religion*.[4] Boyer began his book with a list of religious beliefs and practices that seemed doubly baffling. First, they had no evidential basis, so how could anyone believe them? Second, they seemed to produce no benefits and often appeared costly, so why would anyone *want* to believe them? According to both Boyer and Atran, the main challenge of explaining religion was to explain how beliefs and practices that are both irrational and costly can nevertheless evolve.

A few years later, the New Atheism movement launched with four books: *The God Delusion* by Richard Dawkins, *Breaking the Spell: Religion as Natural Phenomenon* by Daniel Dennett, *God Is Not Great: How Religion Poisons Everything* by Christopher Hitchens, and *The End of Faith: Religion, Terror, and the Future of Reason* by Sam Harris.[5] These books attracted much more attention than the scholarly works of Atran, Boyer, and mine. The New Atheists wanted everyone to know that God does not exist. They also wanted everyone to know that religion as a human construction is detrimental to humankind, which is evident from their book titles. Dawkins is an iconic evolutionist for the general public. Dennett is a philosopher by training but became an intellectual spokesperson for evolution with his book *Darwin's Dangerous Idea: Evolution and the Meanings of Life*.[6] Hitchens and Harris wrapped their arguments in the mantle of science and rationality, even if they did not have specific expertise in evolution.

In short, at the dawn of the twenty-first century, the handful of people writing on religion from an evolutionary perspective displayed the full spectrum of opinion on the impact of religion on human welfare, from "religion is the scourge of humanity" (Dawkins), to "religious beliefs and practices are nonadaptive byproducts" (Atran and Boyer), to "religious groups are like superorganisms" (myself). In some respects we were continuing a debate that goes back to the first scholars who attempted to explain religion as a natural phenomenon, such as Edward Burnett Tyler, James George Frazer, and Emile Durkheim.[7] Some interpreted religion as erroneous attempts to understand the world that would be replaced by superior scientific knowledge. But Durkheim, who was the son of a rabbi, claimed that religions have great "secular utility," as he put it, which led to his well-known definition of religion as "a unified system of beliefs and practices relative to sacred things . . . which unite into one single moral community, called a church, all those who adhere to them."

The twenty-first-century evolutionists were carrying on the same debate, but we were also adding something new. Evolutionary biologists have a well-developed routine for determining whether any given evolved trait is an adaptation. Adaptations require a process of selection. Group-level adaptations require a process of between-group selection for the most part. If a trait is not an adaptation, it can persist in a population for a number of reasons. It can be a product of genetic drift. It can be a byproduct of an adaptation, or a "spandrel," to use an architectural metaphor made famous by Stephen Jay Gould and Richard Lewontin in a classic 1979 article.[8] It can be an adaptation to a past environment that has become mismatched to the current environment. Or it can be an adaptation at a lower level of a multitier hierarchy that is maladaptive at a higher level. All of these possibilities exist for products of cultural evolution in addition to products of genetic evolution. Empirical research is required to test among them, and it is possible to determine the facts of the matter, although it isn't always easy. For example, we know with near certainty that the differences among finch species on the Galapagos Islands are caused largely by selection, even though the prevailing view during the first half of the twentieth century was that they were caused primarily by genetic drift.[9]

What's new about approaching religion from an evolutionary perspective is that progress has been made in discriminating among the alternative hypotheses. There is much more agreement about cultural group selection as an important evolutionary force, resulting in religious systems that organize communities of religious believers, much as Durkheim posited. An especially important advance has been to test adaptation vs. byproduct hypotheses separately for genetic and cultural evolution. The growing consensus is that some elements of religion qualify as byproducts with respect to genetic evolution (e.g., an innate tendency to attribute agency to events, which evolved by genetic evolution without reference to religion) but adaptations with respect to cultural evolution (e.g., particular conceptions of gods as agents that have the effect of motivating group-advantageous behaviors). A sample of recent books that convey this consensus includes *The Evolution of God* by Robert Wright, *The Faith Instinct: How Religion Evolved and Why It Endures* by Nicholas Wade, *Big Gods: How Religion Transformed Cooperation and Conflict* by Ara Norenzayan, and the magisterial *Religion in Human Evolution: From the Paleolithic to the Axial Age* by the renowned sociologist Robert Bellah.[10]

This consensus enables me to make a central claim for the purpose of this book: *Most enduring religions promote altruism expressed among members of the religious community, defined in terms of of action.*[11] In other words, religions cause people to behave for the good of the group and to avoid self-serving behaviors at the expense of other members of their group. This is one of those statements that probably appears obvious in retrospect (if it doesn't already), even though theorists and scholars of religion have contested it for well over a century.

In chapter 2, I distinguished between high-cost behaviors that anyone would call altruistic based on action (such as a soldier falling on a grenade to save his comrades) and low-cost behaviors that benefit the group but which most people would not regard as altruistic (such as coordinating behaviors to make a good collective decision). I concluded that these behaviors are different in degree but not kind because in both cases they do not maximize relative fitness within groups and require a process of group-level selection to evolve. Religions afford plentiful examples of both kinds of adaptations. Hutterites give all their belongings to the church, and many were put to death rather than recanting their beliefs during the early history of the sect. As an example, Klaus Felbinger was a Hutterite and blacksmith by trade who was captured in 1650 by Bavarian authorities and tortured by the Catholic clergy for over ten weeks to make him recant before being executed. His testimony was so eloquent that even his captors were impressed. It survived his death and entered the canon of Hutterite texts.[12]

As an example of low-cost altruism, consider a social convention that Hutterites employ when their colonies fission to form new colonies. Moving to a new site typically involves more hardship than staying at the old site. To avoid disruptive competition within the group, the colony is divided into two lists, and one list is drawn at random to determine who moves. The Hutterites could rely on voluntary altruism during this recurrent phase of their life cycle but they don't, and one can well imagine the history of disputes and hard feelings that led to the adoption of a social convention that guarantees fairness. In general, group-level selection always favors low-cost forms of altruism over high-cost forms, to the point where many of the mechanisms that cause religious groups to function well don't appear altruistic at all. High-cost forms of altruism exist primarily when there is no other way of benefitting the group.

As a second example of low-cost altruism, it is a naïve rendering of "turn the other cheek" to think that Christians are instructed to forgive all trespasses. If they did, then there would be no protection against self-seeking behaviors within groups. The Hutterite blacksmith captured by the Austrian authorities described an explicit procedure for dealing with transgressions that illustrates the design principle of graduated sanctions exhibited by the common-pool resources (CPR) groups studied by Elinor Ostrom. The Hutterite procedure (drawing upon older Christian traditions) begins with friendly "brotherly correction" and escalates to exclusion when necessary, but always with the hope that the person will repent and return to the community, when all will be forgiven. The prohibited behavior, not the person, is being excluded. Most enduring religions bristle with mechanisms of this sort that suppress self-serving behaviors within groups, causing between-group selection to be the dominant cultural evolutionary force. The mechanisms might not seem altruistic, but they are when altruism is defined in terms of traits that benefit whole groups and require between-group selection to evolve. Readers who doubt that altruism at the level of action exists within religious groups are encouraged to read the aforementioned books and additional material referenced in the endnotes of this book.

Now I turn my attention to altruism at the level of thoughts and feelings. Are religious believers motivated to help others as an end in itself, knowing that this will often require a personal cost? The answer to this question appears to be no—and it is surprisingly definitive.

Jacob Neusner and Bruce Chilton are two highly respected theologians and religious scholars at Bard College's Institute for Advanced Theology. The John Templeton Foundation commissioned them to organize a conference on altruism in world religions, resulting in an edited volume that was published in 2005, as I briefly recounted in the Introduction.[13] They asked William Scott Green—another distinguished religious scholar, who coedited the *HarperCollins Dictionary of Religion*—to create a definition of altruism and set of questions that the rest of the of the conference participants, each a specialist on a different world religion or philosophical system, could address. Green's definition, which was also stated at the beginning of this book, is: "Intentional action intended ultimately for the welfare of others that entails at least the possibility of either no benefit or loss to the actor." The questions to be addressed by the conference participants included the following::

> ► Major categories of behavior for the welfare of others.
>
> ► What a given religion means by "others," both doctrinally and historically.
>
> ► Whether historical practice follows the guidance of the text.
>
> ► How the religion assesses the meaning of behavior for the welfare of others.
>
> ► Whether intention or motivation is a factor in determining an action's meaning.
>
> ► Whether the religion creates a context in which benefitting others requires a cost for the actor.

I don't know what John Templeton and others at the foundation had in mind when they commissioned this project. Perhaps they thought that altruism would be the beating heart of all world religions. If so, then they were to be disappointed. Scholar after scholar who followed Green's rubric reported that his definition of altruism did not fit the religious or philosophical system that they were reporting upon. People indeed behaved for the good of their groups, sometimes at great cost to themselves. They just didn't conceptualize it in terms of altruism.

According to Robert M. Berchman, a specialist in Greco-Roman philosophy, "There is little to suggest that contemporary definitions and theories of altruism summarized by William Scott Green . . . fit Greco-Roman definitions of 'selfless activity' at all." Aristotle, for example, distinguished between three kinds of friendship, based on advantage, pleasure, and goodness. The first two might seem to match our concept of selfishness and the third to match our concept of altruism, but Aristotle argues that even the third kind of friendship promotes one's own good. He thought that the self is naturally social so that a concern for others is part of self-interest. Caring for others in the way we care about ourselves does not interfere with our interests but expands them.

Moving from Greco-Roman thought to Judaism, Neusner and Alan J. Avery-Peck conclude that by any rigorous definition of altruism—which means anything beyond philanthropy in a broad sense—altruism is absent from Judaic thought. If it features importantly in Christian thought, then that would be for reasons particular to the theological narrative of Christianity.

But Christianity doesn't live up to Green's definition of altruism any more than Judaism, according to Chilton's contribution to the volume. Jesus regarded the greatest two commandments to be (1) love God and (2) love one's neighbor as oneself, but followers of these commandments were rewarded with eternal life. Where is the altruism in that?

The Hutterites provide an example of a Christian belief system that elicits highly altruistic behavior but does not draw upon altruistic motives, as strange as this might seem. In 1995, while I was still dabbling in religious studies and came upon that Hutterite text, I categorized every phrase that conveyed information about the effects of actions on the welfare of self and others, resulting in a two-by-two table (see Figure 6.1).[15] Altruism as defined by Green would occupy the top left quadrant of the table (negative for self, positive for others), but this and the lower right quadrant (positive for self, negative for others) were empty. In other words, according to the worldview expressed in the text, *all* actions are either beneficial for both self and others (top right quadrant) or detrimental for both self and others (bottom left quadrant). Altruism according to Green's definition truly does not exist within this worldview!

As for Greco-Roman philosophy, Judaism, and Christianity, so also for Islam, classical Buddhism, contemporary Buddhism, classical Hinduism, and Chinese religions, as the other chapters of the volume edited by Neusner and Chilton attest. In an epilogue, Green summarizes the results in the volume this way:

> To many people in contemporary American life, altruism—the ideal of selfless concern and care for the other—appears to be a distinctively religious value. Because all religions teach the importance of caring for others in one fashion or another, it seems to follow that religion should be the natural soil in which the seeds of human altruism can grow.... Our common purpose was to see if

altruism is a useful and appropriate category for the academic study of religion. . . . In typical academic fashion, the answer seems to be both no and yes: "no" because there is broad agreement among the contributors to this volume that the contemporary understanding of altruism used for this experiment either does not apply or is otherwise unsuited to the classical materials of the religions under study; "yes" because, in the process of deciding the "no," the chapters in this anthology collectively reveal the resources for benevolence, charity, and human caring in the foundation texts of the religions they study.

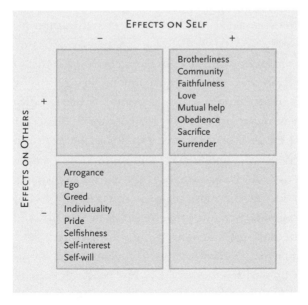

**FIGURE 6.1** THE HUTTERITE WORLDVIEW

Green reaches this conclusion without any reference to evolution, but it perfectly illustrates the one-to-many relationship between ultimate and proximate causation that I stressed in the previous chapter. Most enduring religions are *functionally* similar in their ability to create highly motivated and well-organized groups—but the proximate mechanisms that evolve in any particular case are hugely variable. This was true for the CPR groups that Elinor Ostrom studied, and it is equally true for religious and philosophical traditions that have culturally evolved over much greater time periods. There are many ways to skin a cat, as there are many ways to activate benevolence, charity, and human caring. There is no reason to privilege one proximate mechanism over another, except in terms of its effects on action.

Granted that altruism as defined by Green is only one way to motivate benevolence, but why is it so notably *absent* from religious worldviews? There are good functional reasons to expect highly motivating worldviews to resemble the table that I derived for the Hutterites, with two quadrants richly described and two that are empty. Actions that benefit oneself but not others (the lower right quadrant) or others but not oneself (upper left quadrant) are messy. Work is required to decide whether to perform such actions. Far simpler and more motivating is to classify all actions as win-win or lose-lose, leaving no uncertainty about what to do. These worldviews are false as factual descriptions of the world, since actual actions frequently benefit oneself but not others or others but not oneself. But worldviews are not selected for their factual accuracy, least of all religious worldviews. They are selected on the basis of what they cause people to do.

An adaptive worldview has two requirements. First, it must be highly motivating psychologically. Ideally, it should cause people to rise out of bed every morning brimming with purpose. Second, the actions motivated by the worldview must outcompete the actions motivated by other worldviews. Altruism as defined by Green is probably absent in most religious worldviews because it performs poorly on the first requirement, compared to worldviews that classify everything as win-win or lose-lose.

Even if a worldview is highly motivating, it can still fail if it motivates the wrong actions. A good example is provided by Seventh-Day Adventism, one of the most successful modern Protestant denominations.[17] It is descended from another religious movement called Millerism that no longer exists. William Miller was a farmer and Baptist lay preacher from northern New York State who became convinced in 1833 that the second coming of Jesus would take place in 1843. Belief in imminent return has fueled the zeal of Christians since the first apostles. Usually it is linked to actions that sustain the church, which are thought to be necessary to prepare for the return. Miller's claim was propagated by the print media, which was just coming into its own in America, and went viral in the same way that computer memes go viral today. Thousands of people around the world made ill-advised decisions in preparation for the coming, only to be disappointed. Most believers went back to their earlier, more serviceable religions, but a few couldn't let go. Like so many cultural mutations, they invented new beliefs about why Miller's prediction had failed and what must be done to prepare for Christ's still-imminent return. One of these splinter groups was inspired by a prophetess named Ellen G. White, who claimed to have actually visited heaven and could describe it firsthand. White's constellation of beliefs became Seventh-Day Adventism, which harnesses the motivating power of belief in imminent return to practices on earth that grow the faith. Not only do Adventists famously keep themselves healthy, but there will be a jewel in their crown in heaven for every person whom they save. White actually saw the crowns on the shelf. If Christ hasn't returned by now, it might be your fault for not preparing hard enough.

That's why Seventh-Day Adventism is currently one of today's fastest-growing religions and often does a better job creating schools and hospitals in underdeveloped countries (in addition to churches) than many governments and secular nongovernmental organizations. Mormonism has a similar story—another religion that originated in New York State in the nineteenth century. It was such a new configuration of beliefs that it barely qualified as Christian, yet the altruism at the level of action that Mormonism inspired was spectacular. In what other westward migration did the first wave of pioneers plant crops to be harvested by the second wave?[18]

The actions inspired by these and other religions qualify as altruistic and "for the good of the group" only in a narrow evolutionary sense. They outcompeted other religious beliefs and practices in their vicinity. Notably, in both of these cases, cultural group selection took place without warfare, which was also true for the spread of early Christianity. Of course, we know that in other cases cultural group selection does take place through direct intergroup conflict and that religions can become well adapted to wage war.[19] That might seem hypocritical from some perspectives but should only be expected from a multilevel evolutionary perspective. If we want to solve the most pressing problems of our age, such as world peace and global environmental sustainability, then more cultural evolution is required and it must be guided by a sophisticated knowledge of evolution, as I explain further in the next chapter.

In this chapter, I have evaluated altruism at the level of action and at the level of thoughts and feelings for religious groups. This task might seem impossibly large for a short chapter, but the results are surprisingly clear. Most enduring religions are impressively designed to motivate altruism at the level of action by promoting behaviors that are for the good of the group and suppressing disruptive self-serving behaviors within the group. If that wasn't obvious before (and for some people it wasn't), it should be now. Yet religions typically do not draw upon altruism at the level of thoughts and feelings to motivate altruistic actions, as strange as this might seem. One reason might be that altruism is a messy category to think about because it inherently pits self- and other-regarding preferences against each other. A more motivating approach is to portray all actions as win-win or lose-lose. This portrayal is false as a factual description of the world, but religious worldviews depart from factual reality in so many ways that we shouldn't be surprised when they do so in this particular respect.

This discussion brings us to the coining of the word *altruism* by Auguste Comte in the mid-1800s.[20] Comte was part of a movement to create a moral system that did not require a belief in God, which he called a "Religion of Humanity." The authority of the new religion would be based on science, and Comte played a large role in formulating the modern concept of science, in addition to creating the word *altruism*. Comte held that all knowledge passes through three states: the theological, metaphysical, and "positive," or scientific. He also arranged scientific knowledge in a hierarchy that ascended from mathematics and physics to biology and sociology, which he called "social physics," setting the stage for one influential branch of economic theory, as I describe in the next chapter.

At the time—and to a large extent still—religious disbelief

was associated with decadence, political unrest, and immorality. Comte needed the concept of altruism to capture the moral high ground. Here is how Thomas Dixon, a scholar of the period, described Comte's views:

> The "great problem of human life" for Comte was how to organize society so that egoism would be subordinated to altruism. The aim of the Religion of Humanity was to solve this problem through social organization and individual religious devotions. Comte also sought to tackle the problem using the tools of positive science. Two of the sciences to which Comte turned in this attempt were the biology of animal behavior and the new cerebral science of phrenology. Both of these were used to argue for the naturalness of altruism. Comte went so far as to claim that the discovery of the innateness of the altruistic sentiments ranked alongside the discovery of the motion of the earth as one of the two most important results of modern science.[21]

Comte regarded this view as a direct contradiction of Catholic doctrine, which taught that human nature was entirely sinful and that love of others was only available through divine grace. He thought that what theologians described as the struggle between the law of the flesh and the law of God could be replaced by the scientific distinction between egoistic instincts located in the posterior part of the brain and altruistic instincts located in the anterior part of the brain, based on the phrenological research of Dr. Franz Joseph Gall. Happiness and progress were therefore a matter of promoting universal altruism through a purely scientific religion of humanity.

Dixon continues:

> A concerted effort was under way to dissociate unbelief from decadence, immorality, and revolutionary politics. The new self-sacrificing idea of altruism fit the bill perfectly—it was a virtue good enough for any respectable believer, and one which was, at least to some extent, independent of the language and metaphysics of Christianity. It was claimed by some that this sort of unbelieving altruism was, in fact, morally superior to Christianity, which was an essentially selfish system based on each individual's desire for infinite reward and fear of eternal punishment.[22]

It is ironic for anyone who expects to find altruism (defined in terms of thoughts and feelings) in religion that Comte coined the word because he knew it was not a core part of Catholic theology. Also ironic is that Comte's effort to create a religion of humanity based on the concept of altruism, with himself as the pope, failed miserably. It was not a good way to skin a cat as far as motivating altruism at the level of action is concerned.

I share the desire of Comte and contemporary humanists, secularists, and rationalists to create a worldview that functions well in a practical sense while fully respecting factual reality. Nevertheless, I admire a good religion for its ability to motivate altruism at the level of action, in the same way that I admire the wisdom of a multicellular organism or a beehive. I hope that my intellectual brethren and I can do as well.

## CHAPTER 7

# Altruism and Economics

I N 2003, one year after the publication of *Darwin's Cathedral*, I decided to become part of something larger than myself. Until then, I had functioned primarily as an individual professor and scientist—teaching my courses; doing my research; writing my grants, articles, and books with collaborators of my choice; and traveling the world giving seminars. Of course, this was only possible thanks to an infrastructure of universities, funding agencies, and publishers, not to mention modern societies that make such things as airplanes and hotels possible. In this sense, I was already a tiny ant within a vast superorganism that ran without me worrying about its welfare.

My decision to become part of something larger than myself was more local and intentional. I wanted my university to be the first to teach evolution as a theory that applies to all aspects of humanity in addition to the rest of life. That effort led to EvoS (for *Evo*lutionary *S*tudies, pronounced as one word), the first campuswide program of its kind, as I recount in my book *Evolution for Everyone: How Darwin's Theory Can Change the Way We Think about Our Lives*, which was published in 2007.[1]

Shortly after *Evolution for Everyone* was published I received an email from a man named Jerry Lieberman, who introduced himself as a humanist, making him an intellectual descendent of Auguste Comte. Jerry wanted to start a humanist think tank and was persuaded by my book that it should be informed by evolutionary theory. Here was a second opportunity to create something larger than myself, and I leaped at the opportunity so fast that Jerry must have been surprised. Within the year we had our first major benefactor. His name was Bernard Winograd, and he was the chief operating officer of the Prudential Insurance Corporation's North American operations. He was paid handsomely for managing billions of dollars in assets and had a lifelong interest in evolution. Our fledgling think tank appeared to be on its way!

Then the financial crisis of 2008 struck. Bernard emailed us to say that he could no longer attend to the Evolution Institute, but he was curious to know over the longer term what evolution might have to say about the crisis. I felt like the gentle hobbit Frodo in J. R. R. Tolkien's *Lord of the Rings* trilogy, being sent on a mission by the great wizard Gandolf before departing to fight his own battles.[2] Plucking up my courage, I wrote a proposal to the National Science Foundation's National Evolutionary Synthesis Center (NESCent) to organize a conference titled "The Nature of Regulation: How Evolutionary Theory Can Inform the Regulation of Large-Scale Human Society." The proposal was funded and thus began my education in economics, the subject of this chapter.

The conference led to a series of NESCent-funded workshops, which in turn led to the publication of a special issue of the *Journal of Economic and Behavior Organization* (*JEBO*) in 2013 titled "Evolution as a General Theoretical Framework for Economics and Public Policy," which I coedited with economists John Gowdy and J. Barkley Rosser.[3] That, in turn, led to a major conference titled "Complexity and Evolution: A New Synthesis for Economics" funded by Germany's Ernst Strungmann Forum, which will be held in Frankfurt in early 2015, about the time this book is published. Thanks to these activities, I feel as qualified to report on the subject of economics as on the subject of religion, based not only on my own contributions but on my role as coordinator for dozens of colleagues in economics and allied disciplines. The journey has been memorable, and parts of it are as strange as Frodo's journey to Mordor or a fundamentalist religion to a nonbeliever.[4]

An economy is the production and consumption of goods and services in a human society. All human groups have economies, insofar as their members do things for each other. The earliest archaeological sites for our species typically have objects and materials from distant sites, suggesting that trade networks extend far back in our history. By the time Adam Smith wrote his two great books, *A Theory of Moral Sentiments* and *The Wealth of Nations*,[5] materials from all over the world came together in the production of a woolen coat, and division of labor had become so specialized that several people were required to make a pin. Modern economies have become mind-bogglingly complex. We truly live in a world of our own making, yet to a large extent it is also a world that no one intended.

The idea that a society can run itself without anyone having the welfare of the society in mind is one of the central themes of economics. This idea is also a major challenge to the concept of altruism because, if true, it suggests that altruism defined in terms of thoughts and feelings need not and perhaps even should not exist. If societies run better on the basis of market forces driven by financial self-interest than on the basis of goodwill, then so much the worse for goodwill. Economists such as Friedrich von Hayek and Milton Friedman made this claim, which became public policy through the influence of think tanks such as the Heritage

Foundation and politicians such as Ronald Reagan in the United States and Margaret Thatcher in the United Kingdom.[6] Ayn Rand gave the notion philosophical and artistic expression; she remains a household name over thirty years after her death and nearly sixty years after the publication of her most influential work, the novel *Atlas Shrugged*.[7] In the utopian community founded by John Galt, the central character of *Atlas Shrugged*, the word "give" is banned from the vocabulary of its members.

The history of a line of thought often depends critically on its starting point. When smart people take a wrong turn at the beginning, they often go a long way before realizing their mistake. In fact, sometimes they *never* realize their mistake and charge ever forward, convinced that they are on the right path. It is therefore worth tracing the concept of a society running on the basis of individual greed to its starting point. One of the earliest influential expressions was a satirical poem titled "Fable of the Bees, or Private Vices, Public Benefits," published in 1705 by the Dutch philosopher Bernard Mandeville, which portrayed human society as a fanciful beehive in which all of the bees are motivated by personal greed.[8] The industrious solid citizen was no different from the knave. Mandeville's fable was regarded as a scandalous attack on Christian virtues at the time but became a precursor of modern economic thought. For our purposes, we can note that real bees aren't motivated by anything that remotely resembles personal greed. Mandeville can be forgiven for not being a bee biologist, but metaphors are powerful and this one set the field of economics on the wrong path. One way to recover is by asking what human societies and insect colonies such as beehives actually share in common, as we did in chapter 4 and will revisit later in this chapter.

Smith created another powerful metaphor when he observed

that buyers and sellers in a market economy are guided "as if by an invisible hand" to benefit the common good, even though they primarily have their own financial interests in mind. Smith's invisible hand metaphor became the banner of contemporary free marketers, but it does not fully represent his own views. One thing I have learned is that popular discourse on the economy is almost totally disconnected from serious academic discourse. Smith used the metaphor only three times in the entire corpus of his work. The following passage from a 2005 article titled "Adam Smith and Greed," published by economist Jonathan B. Wright in the *Journal of Private Enterprise*, describes what any good scholar of Smith knows:

> Ethical egoism may be flourishing in American culture, but the association of Adam Smith with these views is simply wrong. Smith decried selfishness often and at length. . . . Smith drew sharp distinctions between greed and selfishness on the one hand and prudent (and virtuous) self-interest on the other. The confusion about Smith's view arises in part from the fact that modern economists put man into the psychological box of *Homo economicus*—an isolated, rational, calculating materialist with no social or moral connections with other human beings, and no scope for heroism. By contrast, Smith found man to be a fundamentally social animal with at times weak powers of rationality and a great capacity for heroic action.[9]

What is *Homo economicus*, and where does this mythical species come from, if not from Smith? That story begins with Leon Walras (1834–1910), a French economist and mathematician whose general equilibrium model laid the groundwork for modern neo-classical economics. Recall from the last chapter that Auguste Comte arranged scientific knowledge in a hierarchy that ascended from mathematics and physics to biology and sociology, which he called "social physics." Physics had been placed on a mathematical foundation by the likes of Isaac Newton and his laws of motion. Walras aspired to do the same for economics. Eric Beinhocker, who currently directs the University of Oxford's Institute for New Economic Thinking, describes the zeitgeist of the times in his book *The Origin of Wealth: Evolution, Complexity, and the Radical Remaking of Economics*:

> Following Newton's monumental discoveries in the seventeenth century, a series of scientists and mathematicians, including Leibniz, Lagrange, Euler, and Hamilton, developed a new mathematical language using differential equations to describe a staggeringly broad range of natural phenomena. Problems that had baffled humankind since the ancient Greeks, from the motions of planets to the vibrations of violin strings, were suddenly mastered. The success of these theories gave scientists a boundless optimism that they could describe any aspect of nature in their equations. Walras and his compatriots were convinced that if the equations of differential calculus could capture the motions of planets and atoms in the universe, these same mathematical techniques could also capture the motion of human minds in the economy.[10]

Walras set about defining people as a kind of atom with fixed preferences and abilities. When the atoms interact, they produce economic activity. The main challenge was to provide mathematical proof that individuals attempting to maximize their personal "utilities" also maximize the welfare of the society as a whole. Walras succeeded at this goal, but only by making a large number of simplifying assumptions about human preferences and abilities, which have become known as *Homo economicus*, as if a description of a biological species. Walras also had to assume that economic systems are at mathematical equilibrium—not because this is realistic, but because it was necessary to do the math.

One can easily understand the allure of a "physics of social behavior" to the nineteenth-century imagination. The mathematical edifice that Walras created is also impressive in its own way. But it was a profound wrong turn in the history of economics, and smart people have been continuing in the wrong direction ever since.

Scholars have noted the unreality of *Homo economicus* from the beginning. Thorstein Veblen's incredulous assessment in 1898 is still among the best: "The hedonistic conception of man is that of a lightning calculator of pleasures and pains who oscillates like a homogeneous globule of desire of happiness under the impulse of stimuli that shift him about the area, but leave him intact."[11] Fast-forwarding to the present, behavioral economists Richard Thaler and Cass Sunstein describe *Homo economicus* this way in their book *Nudge*: "If you look at economics textbooks, you will learn that *Homo economicus* can think like Albert Einstein, store as much memory as IBM's Big Blue, and exercise the willpower of Mahatma Gandhi. Really."[12] These and many other critiques come from *within* the field of economics, so you don't need to take the word of an outsider such as myself!

If *Homo economicus* is so far-fetched, why does it play such an outsized role in economic theory and policy? It was a minority position during the first half of the twentieth century but then rose to prominence through the efforts of a group of intellectuals called the Mont Pelerin Society and the think tanks that it spawned. The two most influential members of the Mont Pelerin Society were the Austrian economist Friedrich von Hayek and the American economist Milton Friedman. Along with Adam Smith, these names are most frequently mentioned to justify the unregulated pursuit of self-interest as enlightened social policy. Ironically, they both used evolution to argue their cases, although in somewhat different ways.

In his classic 1953 article "The Methodology of Positive Economics," Friedman admits that the assumptions of neoclassical economics are unrealistic.[13] Nevertheless, he argues that the predictions of the theory can be right even when its assumptions are wrong. He bases his argument on three analogies. First, trees orient their leaves to maximize exposure to the light, but no one thinks that they solve optimization equations. Second, expert pool players act is if they are performing complex calculations when making their shots, when in fact their abilities are based on countless hours of play. Third, firms maximize their profits as if their executives know what they are doing, when in fact they are merely the survivors of a Darwinian struggle in which inefficient firms went extinct.

All three of these analogies invoke evolutionary processes— genetic evolution in the case of trees, trial-by-error learning for the pool players, and cultural evolution for the firms. In essence, Friedman argued that people and firms have been selected to behave as if the assumptions of neoclassical economic theory are correct, even if the proximate mechanisms are different.

Hayek went further than Friedman in developing the thesis that the intelligence of economic systems does not reside in individual actors. Instead, it is a distributed property of the whole system. For this to happen, there must be a process of selection among economic systems. Hayek pioneered the concept of cultural group selection and self-organizing systems far ahead of his time.[14]

If we judge Friedman's and Hayek's arguments from a modern evolutionary perspective, they are woefully lacking. Friedman would be convicted of the worst kind of "just-so" storytelling of the sort that Stephen Jay Gould and Richard Lewontin cautioned against in a classic paper written in 1979.[15] Hayek was on the right track by emphasizing self-organization, distributed intelligence, and cultural group selection, but his account requires major updating and in all likelihood will lead to very different conclusions than currently drawn from his work. Nor did their evolutionary arguments lead to the proper study of economics from an evolutionary perspective. Friedman's article was used primarily as a shield to deflect criticisms of neoclassical economic theory. Hayek's invocation of cultural group selection came at a time when the rejection of group selection was in full force and fell upon deaf ears among biologists and economists alike.

When I began my economics education, an august body of theory seemed to support the notion that "greed is good." "Laissez-faire leads to the common good" was called the *First Fundamental Theorem of Welfare Economics*.[16] The idea that the only responsibility of a corporation is to maximize profits for its shareholders was called *agency theory*. The more I learned, the more I realized that the aura of authority provided by the mathematics was false. A theory is only as good as its assumptions, and economics of the *Homo economicus* variety is as detached from factual reality as any religion. Smart people have galloped off and gone a very long way in the wrong direction.

But wait—there's more! We have yet to explain the power of

Ayn Rand. How did a Russian immigrant armed with nothing but a typewriter earn a place alongside the likes of Smith, Hayek, and Friedman in the pantheon of free-market deities? Why is she read more than almost any other author of her period? Why do some politicians insist that their staffs read her books?[17] Why are there Ayn Rand clubs on college campuses around the world?[18] I think I know the answers to these questions, based on the same method that I used to study the Hutterites described in the previous chapter.

Recall that according to the Hutterite worldview, all actions are either good for everyone or bad for everyone. Messy win-lose and lose-win situations, which require thinking to resolve, don't exist. The worldview is a flagrant departure from the real world, but those who become believers brim with purpose as they travel in their minds confidently toward glory and away from ruin. What actually befalls them depends upon the actions that their faith calls upon them to do, which can be either glorious or ruinous, as we saw with the comparison between Millerism and Seventh-Day Adventism.

Back in 1995, when I was still dabbling in religion and long before EvoS and the Evolution Institute, I analyzed Rand's book of essays titled *The Virtue of Selfishness* in the same way that I analyzed the Hutterites.[19] Every word, phrase, and passage that could be categorized in terms of effects on the welfare of self and others were placed in one of the four quadrants of a two-by-two table. As we can see for a sample shown here (see Figure 7.1), Rand's worldview is just as linear as the Hutterite worldview, although the pursuit of self-interest is portrayed as good for all, and the traditional virtues, along with self-destructive selfish behaviors, are portrayed as bad for all. Rand even states outright in one of her essays that "there are no conflicts of interest among rational men."[20]

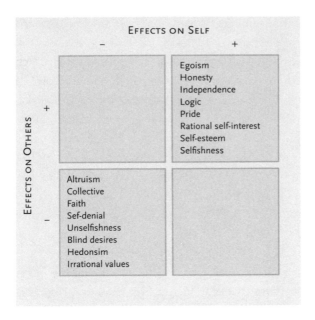

**FIGURE 7.1** AYN RAND'S WORLDVIEW

In short, Rand's worldview adds a layer of unreality on top of the unreality of *Homo economicus*. Those who become believers acquire a kind of zealotry typically associated with religious fundamentalism. Social commentator Maria Bustillos put it well in a 2011 article on Rand's influence: "Rand's books have sold nonstop from the moment they were published because people love hearing how not only can they get away with being totally selfish, it's absolutely the right way to be. The best way to be, as in, morally the best."[21] The following passage by Nathaniel Branden, a disciple of Rand who became her lover before leaving the movement, could have been written by a religious zealot: "This is how we were back then, Ayn and I and all of us—detached from the world—intoxicated by the sensation of flying through the sky in a vision of life that made ordinary existence unendurably dull."[22]

Never mind that Rand was an atheist and clothed her worldview in the language of rationality; the linearity of her worldview exemplifies fundamentalism. Invoking the gods is just a detail. The fundamentalist nature of Rand's thought explains why academic philosophers have never taken her seriously, but she still casts a spell over her readers to this day, not least Alan Greenspan, who was part of her inner circle as a young man and went on to become chair of the U.S. Federal Reserve Board from 1987 to 2006. We should be frightened when politicians insist that their staff read Ayn Rand. The separation of church and state offers no protection against this kind of fundamentalism.

In the previous chapter on religion, I wrote that an adaptive worldview has two requirements. First, it must be highly motivating psychologically. Second, the actions motivated by the worldview must outcompete the actions motivated by other worldviews. But worldviews that are successful in the evolutionary sense of the word need not benefit society as a whole. We know from multilevel selection theory that they can spread, cancerlike, at the *expense* of society as a whole. We are currently witnessing this phenomenon with the spread of the "greed is good" worldview.

Calling a worldview cancerous says nothing about the thoughts or intentions of its believers. To pick Greenspan as an example, by all accounts he is a decent man who thought that he was promoting the common good. During the widely reported congressional hearing that followed the 2008 financial crisis, he was genuinely dumbfounded that the policies he helped to implement led to such disastrous results: "Those of us who have looked to the self-interest of lending institutions to protect shareholders' equity, myself included, are in a state of shocked disbelief."[23] We can sympathize with Greenspan's plight as a good man who caused great harm, which is the stuff of tragedy. But a worldview that causes its believers to do harm while regarding themselves as morally pure is even more insidious than a worldview that actively condones harm.

Given the unreality of *Homo economicus*, coupled with the unreality of Randian fundamentalism, there is an urgent need to start over with the central question of how a society can function well without its members having the welfare of the society in mind. Evolutionary theory can set us on the right path. It turns out that the invisible-hand metaphor can be placed on a strong scientific foundation with important implications for public policy.[24] The endnotes of this book provide a guide to the small but growing academic literature on the subject. For the remainder of this chapter, I make a number of key points that are so elementary from an evolutionary perspective that they are unlikely to be wrong.

The invisible-hand metaphor includes two criteria: (1) a society functions well as a collective unit (2) without the members of the society having its welfare in mind. These two criteria can be examined for nonhuman societies in addition to human societies. As we saw in chapters 1 through 4, nonhuman societies can evolve to function well as collective units, but only when special conditions are met. In all other cases, the first criterion of the invisible-hand metaphor doesn't apply. When nonhuman societies do evolve to function well as collective units, their members typically do not have the welfare of their society in mind, thereby satisfying the second criterion of the invisible-hand metaphor. We can say this with confidence because the members of most nonhuman societies don't even have minds in the human sense of the word.

Consider the genes and cells that constitute a multicellular organism. Calling the organism a society is not stretching the truth, as we saw with the concept of major evolutionary transitions in chapter 3. Genes and cells don't have minds. They simply respond to their local environments in ways that contribute to the organism's survival and reproduction. These responses are a tiny fraction of possible responses, most of which would be detrimental for the organism. The adaptive responses were winnowed from the maladaptive responses by the process of natural selection at the level of the organism. Organisms whose genes and cells behaved inappropriately are not among the ancestors of current organisms.

The same story can be told for social insect colonies as highly functional collective units. Beehives are nothing like what Mandeville imagined. A bee is a highly sophisticated organism in its own right, with a mind according to at least some definitions of the word; when it comes to participating in the economy of the hive, however, the bee responds to local conditions in much the same way as genes and cells, as we saw in the case of collective decisionmaking in chapter 1. The responses that work at the colony level are like needles winnowed from a haystack of maladaptive responses by colony-level selection. Colonies whose members didn't make the right moves are not among the ancestors of current colonies.

In short, nonhuman societies provide outstanding examples of the invisible-hand metaphor when they are products of society-level selection but not otherwise. Higher-level selection *is* the invisible hand. When it operates, lower-level elements behave for the common good without necessarily having its welfare in mind. When higher-level selection doesn't operate, the society ceases to function as a collective unit.

An important difference between the valid concept of the invisible hand and the received economic version is that in the latter case, unrestrained self-interest (usually conceptualized as monetary gain) is thought to robustly benefit the common good. In the former case, unrestrained self-interest is far more likely to *undermine* the common good, either by profiting some members of the society at the expense of others (destructive forms of within-group selection) or by failing to coordinate members of the society in the right way. For every set of responses that work at the society level, countless responses don't work. The idea that group-functional proximate mechanisms are like needles that must be found in a haystack of dysfunctional mechanisms is integral to the valid concept of the invisible hand and largely absent from the received economic view.

Humans have minds, beliefs, and motivations in ways that other species do not, but the functionality of human societies is still a matter of society-level selection, as we saw in chapter 4. Preferences and abilities of the *Homo sapiens* variety, rather than the *Homo economicus* variety, evolved by virtue of causing some groups to survive and reproduce better than other groups. These preferences and abilities include a healthy regard for one's self-interest and resistance to being pushed around by others. They include the ability to establish norms of acceptable behavior by consensus and a willingness to punish norm violations, even when they do not directly impact one's own immediate welfare. Such preferences and abilities include mechanisms such as gossip, ensuring that status is based on reputation and not coercive power. They include a willingness to give freely to others and to the creation of public goods that are straightforwardly "for the good of the group," but only under conditions that merit trust. Astonishingly the need to base economic theory on *Homo sapiens*, not *Homo economicus*, only dawned upon the economics profession in the last quarter of the twentieth century. Parenthetically, the full corpus of Adam Smith's work accords well with modern research. One recent edited volume is titled *Moral Sentiments and Material Interests: The Foundations of Cooperation in Economic Life*, borrowing from the title of Smith's first book.[25]

Some of the mechanisms that cause small groups to function well are based on conscious intentions, but others operate beneath conscious awareness. Like genes, cells, and bees, we play our role without knowing the role we are playing. This gives life in small groups an invisible-hand-like quality that Alexis de Tocqueville, the great French social theorist who visited America in the nineteenth century, noted when he wrote that "the village or township is the only association so perfectly natural that . . . it seems to constitute itself."[26] This is not because social life in small groups is a simple matter, but rather because so much is taking place beneath conscious awareness. By the same token, vision is effortless for us, but the neurobiology of vision is extraordinarily complex. We effortlessly participate in the economy of small groups in the same way that we effortlessly open our eyes and see.

When human societies became larger with the advent of agriculture (and naturally occurring concentrations of resources), the problems of coordination and protection against exploitation from within didn't go away. If anything, they became more demanding. New mechanisms arose by cultural group selection to solve these problems, as we saw in chapter 4. Some were based on the conscious selection of alternatives, but others were the product of raw cultural evolution—many inadvertent social experiments, a few that succeeded. No one truly intended to create the societies that survived. Thus, Smith and others were right to comment on the self-organizing properties of economic and social systems, which do not require and can often be disrupted by conscious planning efforts. But it was a monumental wrong turn to suppose that this kind of self-organization can be derived from a simple notion of individual self-interest, much less the maximization of financial wealth. Centuries and even millennia of cultural evolution were required to winnow the self-organizing processes that work, like needles from the haystack of self-organizing processes that don't work!

Cultural evolution is not a thing of the past. It is operating today faster than ever before, and unless mechanisms of coordination and protection are in place, cultural evolution will lead to outcomes that are highly dysfunctional for society as a whole.[27] An example from today's headlines (as I write these words) is high-frequency trading, the newest innovation of financial markets, as recounted for a general audience by Michael Lewis in his book *Flash Boys: A Wall Street Revolt*.[28] The story begins with a set of rules that the U.S. Securities and Exchange Commission (SEC) passed in 2005 and enacted in 2007, which required brokers to find the best market prices for their investors. The purpose was to prevent trading abuses that had taken place previously, but the rules set the stage for a new kind of abuse. A system for comparing prices across stock markets was required, and the one that the SEC created was slow by computational standards. Traders could exploit this weakness by building faster processors, allowing them to preview the market and trade on what they had seen.

The result was an arms race for processing speed of mind-boggling proportions. *Flash Boys* begins with the story of a fiber optics cable that was built to connect a data center in Chicago to a stock exchange in New Jersey at a cost of tens of millions of dollars, just so that information could be transmitted a few milliseconds faster than existing communication channels. Like the fastest-running cheetah, this tiny difference in speed enabled the owners of the cable to charge banks and investment companies

millions of dollars to gain an edge over their rivals without producing any social benefits whatsoever. This move was just part
of a feeding frenzy in which the high-frequency traders were the
predators and the average investor was the prey. No invisible hand
would save the day. The pursuit of financial gain led to exploitation, pure and simple.

The heroes of *Flash Boys* are a small group of people who discovered the injustice of high-frequency trading and were morally offended by it. In evolutionary terms, they were cooperators
who were inclined to punish wrongdoing. This part of the human
psychological toolkit is required to keep groups of any size on an
even keel, but moralistic punishment only works with a sufficient
number of punishers that can inflict punishment at low cost to
themselves. These conditions are commonly met in small groups,
making moralistic punishment so spontaneous that we scarcely
know it is happening, but the social environment of Wall Street
is wickedly stacked against moralistic punishers. They are few
in number. The cost of attempting to punish wrongdoing can be
severe, and how one can punish isn't clear, even if one decides to
make the sacrifice.

Not everyone who inhabits the world of high-frequency trading is morally corrupt, but the moral landscape is nevertheless
bleak. Some inhabitants are predators and proud of it. Some
know what is taking place but don't see how it can be stopped.
Some enjoy the money and the technological challenges without
thinking or caring about the societal consequences. Some are
misled by free market ideology to think that the system is actually good. Here is how Lewis describes the moral landscape of
high-frequency trading through the eyes of one of the characters
of *Flash Boys*, Don Bollerman:

Don wasn't shocked or even all that disturbed by what had happened, or, if he was, he disguised his feelings. The facts of Wall Street life were inherently brutal, in his view. There was nothing he couldn't imagine someone on Wall Street doing. He was fully aware that the high-frequency traders were preying on investors, and that the exchanges and brokers were being paid to help them to do it. He refused to feel morally outraged or self-righteous about any of it. "I would ask the question, 'On the savannah, are the hyenas and the vultures the bad guys?'" he said. "We have a boom in carcasses on the savannah. So what? It's not their fault. The opportunity is there." To Don's way of thinking, you were never going to change human nature—though you might alter the environment in which it expressed itself.[29]

The biological imagery and invocation of human nature in this passage are instructive. Biological ecosystems are full of interactions among species that would be regarded as immoral in human terms. These interactions exist because the ecosystems are typically not units of selection. The same interactions are largely suppressed in small human groups because groups of a similar size were units of selection during our evolution as a species. Part of our evolved human nature is to regard predatory interactions within our groups as immoral and to try to suppress them. Of course, it is also part of human nature to become predatory when protections are not in place. The current Wall Street environment rewards predatory behavior, but it could equally reward moralistic behavior if suitable protections could be put into place.

Neuroimaging studies show that people who become morally

indignant take pleasure in seeking revenge.[30] The same centers in their brains get activated as for purely personal gratification. The sweetness of revenge might not count as altruistic in terms of thoughts and feelings, but it motivates behaviors that count as altruistic in terms of the time, energy, and risk required to punish wrongdoing. Some of the characters in *Flash Boys* bring these scientific results to life, such as John Schwall, who felt this way about high-frequency trading: "It really just pissed me off," he said. "That people set out this way to make money from everyone else's retirement account. I knew who was being screwed, people like my mom and pop, and I became hell-bent on figuring out who was doing the screwing."[31]

It might seem that every effort to prevent predatory behavior will select for new types of predatory behavior in an endless coevolutionary race. After all, the opportunity for high-frequency trading was made possible by a SEC ruling designed to prevent prior trading abuses. But environments can be constructed to make predation easy or difficult. Plant bushes around a waterhole and it becomes easy for predators to ambush their prey. Remove the bushes and ambushing prey becomes difficult or impossible. The main achievement chronicled in *Flash Boys* is the creation of a new stock market (called Investors Exchange, or IEX) with fairness baked into it, unlike all the other stock markets. For the first time, investors have an opportunity to trade in an environment that is remarkably well protected against predators. At the present moment, a process of cultural evolution is in progress: a Darwinian competition between a fair stock market vs. stock markets that favor predatory activities in myriad forms. If the fair stock market prevails, the impact on societal welfare could be huge, but this outcome requires a moral choice on the part of those in a position to select among stock markets.

How does this chapter address the question of whether altruism exists? One of the greatest challenges to the concept of altruism has been the notion that societies can function well based entirely on individual self-interest. For nearly half a century, economics of the *Homo economicus* variety has seemed to provide authoritative support for a narrow version of this view that equates individual self-interest with the maximization of financial gain, as if anything of value can be given a monetary value and compared to anything else of value through market processes. Fundamentalism of the Ayn Rand variety has been extraordinarily successful at convincing people that the pursuit of self-interest is morally pure and will benefit everyone in the long run. One contribution of this chapter is to show that this paradigm has no legitimate authority. If you remain skeptical, then I encourage you to read the academic literature that stands behind this chapter, as referenced in the endnotes.

Even though the received version of the invisible-hand metaphor can be authoritatively rejected, a more legitimate version exists that can be understood from an evolutionary perspective. Societies function well when they are a product of society-level selection. The proximate mechanisms that evolve need not require having the welfare of the society in mind. In the case of nonhuman societies, the proximate mechanisms don't even require having minds in the human sense of the word.

All societies that function well require mechanisms that coordinate action and prevent exploitation from within. Whether these mechanisms count as altruistic or selfish, defined in terms of thoughts and feelings, is irrelevant, as long as they do the job in terms of actions. A panoply of proximate mechanisms evolved by genetic evolution in our species that keep small groups on an even keel; these mechanisms operate so spontaneously that we tend to forget that they exist. Small groups thus have an invisible-hand-like quality, as Alexis de Tocqueville observed when he wrote that the village or township is the only association so perfectly natural that it seems to constitute itself.

Large societies function well, to the extent that they do, only thanks to proximate mechanisms that evolved by cultural evolution and that interface with our genetically evolved mechanisms. We know about some of these mechanisms, especially those that we intentionally designed in our conscious efforts to improve the welfare of society (e.g., laws and constitutions). But large societies function well in part because of mechanisms that no one planned or intended, but that nevertheless caused some groups to outcompete other groups. This gives large societies an invisible-hand-like quality, as noted by early thinkers such as Bernard Mandeville and Adam Smith, but it was a monumental mistake to conclude that something as complex as a large society can self-organize on the basis of individual greed.

Future social arrangements need to be based more on intentional planning than ever before. This does not necessarily mean *centralized* planning; it can also include the smart design of decentralized processes. Either way, some structure must be imposed to coordinate action and prevent exploitation from within. For this structure to function at the global scale, welfare at the global scale must be the selection criterion. Anything less will result in the dysfunctions inherent in multilevel selection theory. In our role as selection agents, we must function as altruists of the highest order.

## CHAPTER 8

# Altruism in Everyday Life

I N 2006, three years after starting EvoS and a year before the start of the Evolution Institute, I began using my hometown of Binghamton, New York, as a field site for studying people in the context of their everyday lives. Field studies are the backbone of evolutionary research because species must be studied in relation to the environmental forces that shaped their properties. What would we know about chimpanzees without the kind of field research that Jane Goodall made famous? What would we know about Darwin's finches if we didn't study them in the Galapagos Islands?

Most research on human behavior is not like this. The bulk of psychological research is conducted on college students without any reference to their everyday lives. Sociologists and cultural anthropologists conduct field research but seldom from an evolutionary perspective. Then there are dozens of research communities with the practical objective of solving a particular social problem, such as addiction, child abuse, or risky adolescent behavior. This kind of applied research often takes place outside the laboratory, but the research communities tend to be isolated from each other and from the so-called basic scientific disciplines, including evolutionary science.

The idea of studying people from all walks of life as they go about their daily lives from a unified theoretical perspective therefore turned out to be surprisingly novel.[1] EvoS and the Evolution Institute enabled me to expand the approach beyond my own efforts. We call the Binghamton Neighborhood Project a "whole university / whole city" approach to community-based research and action. The "whole university" part is the network of faculty and students organized by EvoS, who speak a common theoretical language in addition to their disciplinary training. The "whole city" part is all sectors of the city—the neighborhoods and their residents; the government; the schools; the health, safety, and social service agencies; and the many other organizations that exist in any city to improve the quality of life. The Evolution Institute has organized workshops on topics such as education, risky adolescent behavior, and quality of life that can be used to implement new policies and practices at any location.[2]

This synthesis enables the study of altruism in the context of everyday life. The Darwinian contest between altruistic and selfish social strategies, defined at the level of action, is taking place all around us. It results not only in genetic and cultural change over the long term but also in the capacities that we develop during the course of our lifetimes and the choices we make on a moment-by-moment basis. Seeing the Darwinian contest clearly can help us create social environments that favor altruism as the winning social strategy, improving the quality of life in anyone's hometown.

Our first project in Binghamton involved, with the help of the Binghamton City School District, giving a survey called the Developmental Assets Profile (DAP) to public school students in grades

6–12. The DAP was developed by a nonprofit organization called Search Institute that has been dedicated to promoting healthy youth development for more than fifty years.[3] The profile measures properties of individuals that contribute to their healthy development (internal assets) and social support from various sources such as family, neighborhood, school, religion, and extracurricular activities (external assets). Among the internal assets are attitudes toward others and society as a whole, measured by statements such as "I think it is important to help other people," "I am sensitive to the needs of others," and "I am trying to make my community a better place." Students indicated their agreement with these statements on a scale from 1 (low) to 5 (high), which we summed to calculate a score that we called PROSOCIAL. The term *prosocial* refers to any attitude, behavior, or institution oriented toward the welfare of others or society as a whole. The term is agnostic about the amount of sacrifice required to help others or to psychological motivation. Thus, it is largely synonymous with what I have been calling altruism as defined in terms of action in this book.

Unsurprisingly, the distribution of PROSOCIAL scores took the form of a bell-shaped curve. Most students indicated moderate agreement with the items. A few whom we dubbed High-PROs appeared to be budding Mother Teresas by indicating strong agreement, and a few that we dubbed Low-PROs appeared to be budding sociopaths by indicating strong disagreement.

So far we had not done anything special. Hundreds of schools and communities use the DAP to measure the developmental assets of their youth. In addition, the PROSOCIAL score doesn't necessarily measure altruism at the level of either actions or thoughts and feelings. Perhaps the High-PROs were just trying to make themselves look good. More work would be required to show that High-PROs walk the walk in addition to talking the talk.

The next thing we did was more novel. By linking the PRO-SOCIAL scores of the students with their residential addresses, we created a map of the city of Binghamton (see Figure 8.1).[4] For those unfamiliar with Binghamton, it is located at the confluence of two rivers, the Susquehanna and Chenango, and the downtown area is between the two arms of the Y. The residential addresses of the students would look like a scatter of points on the map. These points were converted to a continuous surface by a method known as kriging, which calculates an extrapolated PROSOCIAL score for each location on the map by taking an average of the actual PROSOCIAL scores for the students closest to the location. The dark regions represent neighborhoods in which most of the students report being highly prosocial. The light regions represent neighborhoods in which most of the students could care less. The spread among neighborhoods was a whopping 50 points on a 0–100 scale!

This map had a powerful effect upon me. Remember that I am trained as a biologist and I am accustomed to seeing maps like this that show the distribution and abundance of nonhuman species. Suppose that the map shown here was for a native plant species that we wanted to preserve. Our job as conservation biologists would be to identify the environmental conditions that cause the species to be common in some areas and absent in others. Perhaps it is soil nutrients, a competing plant species, or herbivores. Once we identify the salient conditions, we can try to provide them more widely, thereby expanding the plant's niche. If we succeed, then the map will grow darker and darker with each census.

The power of the map is to suggest that a human social strategy such as prosociality could be conceptualized in the same way. Most people are behaviorally flexible and can calibrate their prosociality to their circumstances.[5] The map suggested that the

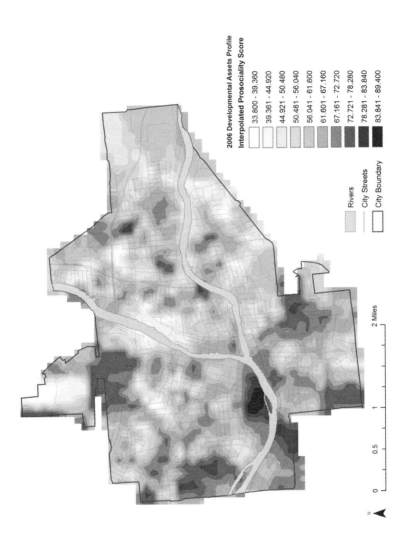

**FIGURE 8.1** A MAP OF PROSOCIALITY FOR THE CITY OF BINGHAMTON, NEW YORK

expression of prosociality in the city of Binghamton was spatially heterogeneous. If the underlying environmental factors could be identified, then the niche for prosociality could be expanded in the same way as the niche for a plant species. If we administered the DAP to the students every year, then we could track our success or failure by watching the dark areas expand or contract.

The school superintendent and teachers who helped administer the DAP shared my excitement about the map and what it represented. They had more practical experience with the students than I ever would, and some of them were born and raised in the Binghamton area. Even for them, the map seemed to show something new and to offer the promise that visualizing the distribution and abundance of prosociality might be the starting point for expanding its niche.

What factors are likely to influence the expression of prosociality in behaviorally flexible individuals? Roughly, they are the same as the factors that favor prosociality as a product of genetic and cultural evolution. High-PROs can expect to succeed when they are in the company of other High-PROs. Otherwise they can expect their communitarian efforts to be exploited and wasted. The other assets measured by the DAP enabled us to test this hypothesis. Bearing in mind that not all social interactions take place within neighborhoods, we correlated the PROSOCIAL score of each individual student with the amount of social support they reported receiving from their family, neighborhood, church, school, and extracurricular activities. The result was a correlation between the prosociality of the *individual* and the overall prosociality of the *individual's social environment* with a coefficient of 0.7 (on a scale from −1 to 1). In plain English, those who reported giving also reported receiving, which is the basic requirement for altruism defined in terms of actions to succeed in a Darwinian world. The results could have been otherwise. The amount of social support that individuals reported receiving could have been uncorrelated or negatively correlated with their own PROSOCIAL score, but this was not the case.

The size of the correlation coefficient is remarkable against the background of evolutionary theories of social behavior. Much has been made of the genetic similarity between relatives as the foundation for the evolution of altruism. The expression of altruism is expected to be proportional to the correlation coefficient for shared genes, which is 1 for identical twins, 0.5 for full siblings, and so on down to 0 for individuals who are genetically unrelated. An important event in the history of inclusive fitness theory was the realization that the correlation between donors and recipients of altruism can be based on factors other than genealogical relatedness, as I recounted in chapter 4. The kind of behavioral sorting that takes place among water striders described in chapter 3 provides an example in nonhuman species. A correlation coefficient of 0.7 suggests that mechanisms *causing those who give to also receive* in the city of Binghamton are so strong that they exceed the correlation among full siblings in a simple genetic model.

A lot of work was required to confirm that the map based on the PROSOCIAL scores actually represented differences among neighborhoods in prosociality.[6] In one study, we employed a method from social psychology that involved dropping stamped addressed envelopes on the sidewalks all over the city to see who was kind enough to put them in the mailbox. In a second study, we employed a method from behavioral economics that involved students choosing between cooperative and noncooperative strategies in a game played for money. In a third study, we measured naturalistic expressions of prosociality such as holiday displays during Halloween and Christmas. In a fourth study, we conducted door-to-door surveys of adult residents. All of the studies broadly supported each other and revealed important new findings. For example, it turns out that the richest kids in Binghamton are not necessarily the most prosocial. Having money means that you don't need to cooperate very much with others to get what you want. In sociological parlance, financial capital and social capital substitute for each other to a certain extent.[7] The most prosocial kids in Binghamton don't have much money but have lots of practice cooperating with each other to get things done. The least prosocial kids in Binghamton have neither financial nor social capital. They have become lone wolves.

Once we were convinced that our map reflected real differences in prosociality, we began to focus on other questions: What are the mechanisms of behavioral flexibility? Can individuals calibrate their degree of prosociality to their social environments throughout their lives? Or do they lose their flexibility as they grow older? There is no single answer. A lot is known about mechanisms of behavioral flexibility in humans and other species. Early childhood experiences can have lifelong consequences.[8] The influence of the environment can take place before birth and even previous generations.[9] The burgeoning field of epigenetics shows that environmental effects can alter the expression of genes in ways that are transmitted across generations.[10] Your propensity to be prosocial is conceivably the result of the experiences of your parents, grandparents, and even more distant ancestors. In addition to environmental effects, genetic polymorphisms cause humans and other species to respond to their environments in different ways. Some individuals are like hardy dandelions that can grow under a wide range of conditions. Others are like orchids that wilt under harsh conditions but grow into objects of rare beauty under optimal conditions.[11]

These genetic and long-lasting environmental effects notwithstanding, many nonhuman organisms retain an impressive ability to respond to their immediate environments. A snail quickly withdraws into its shell when it senses danger and emerges just as quickly when danger passes. Our research suggests that people have a snail-like ability to change their behavior in response to the prosociality of their social environment, regardless of their past experiences.

In one study headed by Daniel T. O'Brien, who was my graduate student at the time and is now doing similar work in Boston, we asked Binghamton University college students (most of whom were unfamiliar with the city) to rate the quality of neighborhoods based on photographs. Their rating correlated quite highly with what the Binghamton public schools students who actually lived in the neighborhoods indicated on the DAP. In a second study, Binghamton University students viewed photographs of the neighborhoods and then played an experimental economics game with public school students from the neighborhoods. This was not an imaginary game but was played for real money. The players did not physically interact, but the college student's decision whether to cooperate in the game was paired with the decision of a public school student from the neighborhood, based on our previous study with the public school students. The college student's willingness to cooperate was strongly correlated with the quality of the neighborhood assessed from the photograph. Just like snails, it appears that we assess our surroundings and withdraw into our shells when we sense danger. In our case, this means not only protecting ourselves from physical harm but also withholding our prosociality. We can transform from a High-PRO to a Low-PRO in the blink of an eye.[12]

A natural experiment that took place in Binghamton enabled us to measure behavioral flexibility another way.[13] Three years after we created the first map, we again administered the DAP. Hundreds of students took the DAP both times, so we could measure changes in their prosociality and other developmental assets over the three-year period. Of these, some had moved their residential location within the city of Binghamton. Their new neighborhood might be better or worse than their old one. Statistically, we were able to test whether the change in their social environment rubbed off on their own prosociality. It did. The students became more or less prosocial depending upon the quality of their new neighborhood. They remained behaviorally flexible, regardless of whatever genetic and long-lasting environmental effects might also be operating.

Natural experiments such as this one are suggestive but seldom definitive. In 2010 we had an opportunity to conduct a more rigorous test when the Binghamton City School District asked us to help them design a new program for at-risk high school students called the Regents Academy. They realized that students who were failing most of their courses in the eighth and ninth grades would almost certainly drop out unless something was done. The school district was willing to create a school within a school, complete with its own principal and staff, to solve the problem. Could we help them design the program?

I leaped at this opportunity, just as I leaped at the opportunity to create the Evolution Institute. One of my graduate students, Richard A. Kauffman, had been a public school teacher before returning to school and joined me on the project. I was also working with Elinor Ostrom, whom I had met as part of my economic work described in the previous chapter. With her postdoctoral associate Michael Cox, we were placing the design principles that she derived for common-pool resource groups on a more general theoretical foundation.[14] The design principles provide a highly favorable social environment for the expression of prosociality. They ensure that efforts on behalf of others and the group as a whole will be well-coordinated (design principles 1 and 3), won't be exploited from within (design principles 2–6), and will be free from external interference (design principles 7–8). (See chapter 1 for more on the design principles.) As such, they provide a blueprint for almost any group whose members must work together to achieve a common goal.

We therefore designed the Regents Academy to embody the eight core design principles. Students, teachers, and staff should have a strong group identity and understanding of purpose. The distribution of work should be equitable, and people should be credited for their contributions. Students should be involved in decisionmaking as much as possible. Good behavior should be monitored, and corrections should be gentle and friendly, escalating only when necessary. Conflict resolution must be fast and regarded as fair by everyone involved. The school must be allowed to create its own social environment as much as possible and to draw upon other groups in constructive ways. In addition to "Ostrom's 8," as we affectionately called them, we added two additional design principles that we thought were especially important in an educational context. The ninth was to create an atmosphere of safety and security, since fear is not an atmosphere conducive to long-term learning. The tenth was to make long-term learning objectives also rewarding over the short term. Research shows that even gifted students succeed primarily when they enjoy developing their talents on a day-to-day basis.[15] Nobody learns when all the costs are in the present and all the benefits are in the far future.

These design principles make perfect sense when they are listed, but how many groups actually employ them? Ostrom found that common-pool resource groups *vary* in their employment of the design principles, which is how she was able to derive them in the first place. The same variation exists for other kinds of groups. As we were designing the Regents Academy, we were reading the educational literature and encountering some school programs that worked very well. Unsurprisingly (to us), they tended to employ the design principles. But the fact that these programs worked well did not mean that they had become common educational practice. On the contrary, education in America (and many other countries) appears to be driven by forces that cause massive violations of the design principles. That was definitely the case for at-risk students in the city of Binghamton. Their experience at the Regents Academy would be very different from their experience at Binghamton's single high school.

To select students to enter the Regents Academy, we compiled a list of all eighth and ninth graders who had failed three or more of their basic subjects during the previous year. Then we randomly selected students from this group to invite to join the Regents Academy. The others formed a comparison group that we tracked as they continued their normal education. Almost all of the families contacted agreed to have their children join the Regents Academy, so we succeeded at creating a randomized control trial, which is the gold standard for studies of this sort. We would know with a high degree of confidence whether the Regents Academy succeeded or failed.

What happened? Within the first marking quarter, the Regents Academy students were performing much better than the comparison group, and the difference was maintained for the rest of the year. The most stringent comparison came at the end of the year when all students took the same state-mandated exams. Not only did the Regents Academy students greatly outperform the comparison group, but they actually performed on a par with the average high school student in the city of Binghamton.[16]

The Regents Academy also improved the developmental assets of the students, as measured by the DAP.[17] Regents Academy students felt better about themselves and reported experiencing more family support than the comparison group. They liked school even better than the average Binghamton high school student, suggesting that all students can benefit from the design principles approach, not just at-risk students. Their PROSOCIAL scores went up, although not enough to be statistically significant. The benefits were achieved to an equal degree by all gender, ethnic, and racial categories. Everyone thrives in a social environment that causes prosociality to win the Darwinian contest.

I have based this chapter primarily on my own experience studying prosociality (= altruism defined at the level of action) in the context of everyday life in my hometown of Binghamton. In addition, through the Evolution Institute I have become familiar with many other programs that employ the same principles and achieve similarly impressive results—not just for education, but for nearly all aspects of everyday life, such as mental health, parenting practices, and the design of thriving neighborhoods. As one example, a way of organizing elementary school classroom activities called the *good behavior game* exemplifies the design principles and has lifelong positive effects that have been demonstrated in randomized control trials.[18] A program that teaches good parenting skills called Triple P (for Positive Parenting Program) originated in Australia and is being implemented worldwide. In one randomized control trial, Triple P was implemented in nine South Carolina counties randomly selected from eighteen counties sharing similar characteristics. Triple P improved parenting practices at a population-wide scale, as measured by public health statistics such as foster care placements and hospital emergency room visits by children. The estimated cost of implementing Triple P in a county was eleven dollars per child, which is trivial compared to the societal benefits, including cost savings by the many social services that must deal with the fallout of bad parenting practices.[19]

Every time I encounter one of these programs, I am amazed at its success in improving human welfare and the quality of the science validating the results, yet I am equally amazed that they aren't more widely known and implemented. Paradoxically, the world is full of programs that work but don't spread. One reason is that they emanate from so many isolated research communities that lack a common theoretical framework.

To address this problem I teamed up with three leading researchers from the applied behavioral sciences, Steven C. Hayes, Anthony Biglan, and Dennis Embry, to write an article titled "Evolving the Future: Toward a Science of Intentional Change."[20] The article has two main objectives: to sketch a basic science of intentional change rooted in evolutionary theory and to review outstanding examples of positive intentional change that are well validated but little known outside their disciplinary boundaries. As we state in our article, we are closer to a science of intentional change than one might think.

Readers should consult the article for a closer look at how to improve the quality of everyday life. I conclude this chapter by making a few points that are most relevant to the topic of altruism. We sometimes need to remind ourselves how much human social life requires prosociality—people acting on behalf of other people and their groups as a whole. The need for prosociality begins before birth and continues throughout development. Most of our personal attributes that we regard as individual because they can be measured in individuals are the result of developmental processes that are highly social. Prosociality during development is a master variable. Having it results in multiple assets. Being deprived of it results in multiple liabilities. This point is often obscured by the fact that every problem experienced by young people tends to be studied in isolation and without a unifying theoretical framework.

After we become adults, almost all of our activities require coordinating our activities and performing services for each other. Some of these activities take place without requiring our attention, as if guided by an invisible hand, but only if there has been a past history of selection favoring self-organizing processes at the appropriate scale, as we saw in the last chapter. At the local scale—in our families, neighborhoods, churches, and businesses—prosocial activities are up to us. If we don't do them, they don't get done and the quality of life declines.

Behaving prosocially does not necessarily require having the welfare of others in mind, as we have seen in previous chapters. Theoretically, Low-PROs could contribute as much to society as High-PROs. A proponent of Ayn Rand might even expect Low-PROs to contribute more to society than well-meaning but counterproductive High-PROs. The results of our research tell a different story. People who agree with statements such as "I think it is important to help other people" actually *do* help other people and work toward common goals more than people who express indifference toward helping other people. For the most part, no invisible hand exists to convert the mentality of Low-PROs into prosocial activities.

Of course, someone who scores high on the PROSOCIALITY scale does not necessarily qualify as an altruist, defined in terms of thoughts and feelings. All sorts of selfish motivations can result in a desire to help others. But obsessing about the motives of High-PROs is like obsessing about whether to be paid by cash or check.

In my experience, people engaged in prosocial activities tend to care about motives only insofar as they bear upon commitment to the task. As long as you are a solid member of the team, you can think about it however you like. The many-to-one relationship between proximate and ultimate causation helps to explain why this pragmatic attitude makes sense.[21]

High-PROs might be essential for volunteer organizations, neighborhood associations, and churches, but how about businesses? There is a widespread impression that businesses are and must be driven by selfish financial motives, but an important book titled *Give and Take*[22] by Adam Grant, who is a professor at the University of Pennsylvania's Wharton Business School, shows otherwise. Grant distinguishes between three broad social strategies. Givers (= High-PROs) freely offer their services without expectation of return. Reciprocators make sure that their services are repaid. Takers (= Low-PROs) try to get as much as they can and give as little as possible in return. It might seem that givers are quickly eliminated in the hardball world of business, leaving only reciprocators and takers, but this is an illusion. Grant reviews an impressive body of evidence showing that givers are highly successful in the business world, as long as they surround themselves with other givers and avoid the depredations of takers. In other words, the business world is no different than any other enterprise requiring prosocial activities.

Why should this be so surprising? How did the impression that businesses must be driven by greed get established in the first place? Most business schools wholeheartedly adopted the *Homo economicus* version of economics described in the previous chapter as their worldview. *Homo economicus* is driven entirely by greed, so that's how we've all been trained to believe regarding the business world. During a video interview that I conducted with Grant, he said that Wharton School students frequently tell him how much they want to help others and society as a whole—which they plan to do by first making a lot of money and then becoming philanthropists! It doesn't occur to them that they can begin functioning as High-PROs right away.

Evolutionary theory and a large body of evidence reveals the importance of being directly motivated to help others and society as a whole—to be a High-PRO and not a Low-PRO. But becoming a High-PRO requires more than counseling and encouragement. It requires building social environments that cause prosociality to succeed in a Darwinian world. Provide such an environment, and people will become High-PROs without needing to be told.

## CHAPTER 9

# Pathological Altruism

LMOST EVERYONE agrees that selfishness comes in benign and pathological forms, which are distinguished by qualifiers such as "enlightened self-interest." When it comes to altruism, however, it is common to think that only good can come from it. I experienced this for myself when I coedited a book titled *Pathological Altruism* with Barbara Oakley, Ariel Knafo, and Guruprasad Madhavan.[1] People who encountered the title before reading the book often acted offended, as if we were committing an act of sacrilege. Had we mysteriously been converted to the creed of Ayn Rand?

Not in the least. The fact that altruism can have both benign and pathological outcomes follows from evolutionary theory at an elementary level. If the concept of pathological altruism clashes with the way that most people think about altruism, then that illustrates the novelty of evolutionary thinking against the background of received wisdom.

Adaptations are seldom categorically good. They contribute to survival and reproduction in some environments, which can be benign, but not others, which can be pathological. Even this statement must be qualified with the word "can," because adaptations in the evolutionary sense of the word are not necessarily benign in the everyday sense of the word. Adaptations can be good for me but bad for you; good for us but bad for them; or good for me, us, and them over the short term but not the long term. These general points apply to altruism as well as any other adaptation.

In chapter 8 I compared the spatial distribution of prosociality in the city of Binghamton to the spatial distribution of a plant species. No one would try to expand the range of the plant species without providing the appropriate growth conditions, but people are urged to behave prosocially all the time without much thought about the conditions that are required to sustain it in a Darwinian world. In some of our Evolution Institute workshops, we have started to call this "declawing the cat."[2] Cats need their claws, especially alley cats, and declawing them only benefits their clawed competitors. People inhabiting harsh social environments similarly need to protect themselves and often have no recourse but to live at the expense of other people. Counseling them to abandon these survival and reproductive strategies without altering their social environment will be either ineffective, if they have enough sense to ignore the advice, or will harm them without providing long-term benefits to anyone else. That's pathological.

I had an opportunity to study the benign and pathological manifestations of altruism in a collaborative project with psychologist Mihaly Csikszentmihalyi, who is best known for his book *Flow*, which is about the experience of being totally immersed in what one is doing. We were able to retrofit a large nationwide study that he and sociologist Barbara Schneider conducted on how American high school students prepare themselves for the workplace.[3] Using the many questionnaire items that the students were asked, we were able to assign a PROSOCIALITY score for each student, much as I did for the Developmental Assets Profile described in the previous chapter. Among the other questionnaire items, the students were asked whether they had experienced a number of stressful events during the previous two years, such as witnessing a violent crime, being assaulted, or being shot at (the sample size was sufficiently large that over 150 students reported being shot at). High-PROs were less likely to experience these events than Low-PROs, which makes sense, given what we have learned about High-PROs clustering together. When High-PROs *did* experience these events, however, they reported feeling more stressed than low-PROs who experienced the same events. Their "niche" was to avoid such tactics altogether, and they were poorly adapted to "niches" where such tactics were commonly employed.

*Pathological Altruism* provides an inventory of psychological dysfunctions and material hazards that high-PROs are likely to experience when they find themselves outside their niche. High-PROs are prone to feeling guilty at the misfortune of others, even when it is not their fault, leading to depression, posttraumatic stress disorder, and obsessive-compulsion disorder.[4] Empathy can trap people in codependent relationships that are highly dysfunctional.[5] Eating disorders can result as much from a desire to sacrifice one's own needs for the sake of others as from a desire to conform to cultural standards of beauty.[6] Even the most prosocial person can experience burnout in high-need situations, such as caring for terminally ill patients.[7]

One of the most detailed and nuanced analyses of pathological altruism comes from Barbara Oakley, lead editor of the *Pathological Altruism* volume, in a book titled *Cold-Blooded Kindness*.[8] Just as murder has been used to reflect upon the human condition in great works of literature such as Shakespeare's *Macbeth* and Truman Capote's *In Cold Blood*, Barbara brought the scientific knowledge that we were assembling in *Pathological Altruism* to bear upon a murder committed by an artist, animal lover, and mother of five named Carole Alden, who phoned the police to say she had killed her husband in self-defense. Oakley was drawn to study this particular case because it seemed like a classic example of spousal abuse. According to the press reports, Alden was a woman who was too kind for her own good, whose need to help others was so strong that she became easy prey for a social predator. This would be one pathological outcome of altruism, but when Oakley dug into the case, she discovered something more complex, dark, and interesting. In many respects, Alden had turned victimhood into an art form. She was the predator and the man she killed was not her only victim.

The actual facts of the case illustrate another pathological outcome of altruism—codependency—in which the desire to help others contributes to a mutually destructive relationship, even ending in a death spiral. There is no clear victim or victimizer in a mutually destructive relationship because each person plays a role in its maintenance. It is not as simple as a predator grasping a prey that is struggling to get away.

Another pathology detailed by Oakley in *Cold-Blooded Kindness* is the murder of objective truth-seeking by passionate advocacy. One might think that an issue as important as spousal abuse would receive the most careful attention that science has to offer, but the crusade to raise consciousness about the issue has led to a black-and-white view of victims and victimizers that does not admit shades of gray. Thus, the authentic scientific study of mutually destructive relationships is only beginning and much remains to be learned. Oakley's inquiry stands in refreshing contrast to well-meaning but misguided crusaders.

Finally, it might seem that a scientific diagnosis of a murder would lack compassion, but nothing could be further from the truth. The picture that emerges from *Cold-Blooded Kindness* is that everyone is trying to survive as best they can, surrounded by kith and kin who love them no matter how badly they stumble. Anyone can become lost in a maze of unforeseen consequences, high-PROs and low-PROs alike. Some of those most closely associated with the case have thanked Oakley for helping them achieve a kind of closure that had previously eluded them.

Even when altruism and other forms of prosociality work as intended, resulting in efficacious groups whose members provide benefits for each other, these groups can do harm to other groups or to the larger society within which they are embedded. The iron law of multilevel selection is that adaptation at any given level of a multitier hierarchy requires a process of selection at that level and tends to be undermined by selection at lower levels. In other words, everything pathological associated with individual selfishness also characterizes altruism at intermediate levels of a multitier hierarchy. My individual selfishness can be bad for my family, but family-level altruism can be bad for the clan, clan-level altruism can be bad for the nation, and nation-level altruism can be bad for the global village.

Small groups have an especially strong grip on our allegiance, which makes sense given our long evolutionary history in small groups. In a study of Machiavellianism, an axis of individual difference related to prosociality, subjects performed a task with a partner who was a confederate of the experiment and who encouraged the subject to collaborate in cheating. The participants could resist cheating at a variety of levels: by attempting to stop their partner, by terminating the experiment, by reporting their partner to the experimenter, or by quickly confessing when the experimenter became suspicious at the end of the experiment. In fact, nearly all participants allowed themselves to become implicated in the unethical act, regardless of their degree of Machiavellianism. The impulse to collaborate with one's immediate social partner— even in an ephemeral social interaction with a stranger—trumped higher-level considerations.[9] The same phenomenon is observed in children who refuse to tell on each other and reserve a special form of contempt for those who do.

In a classic longitudinal study of deviance, thirteen-year-old boys were videotaped while talking with a friend.[10] Laughing about deviant behavior was highly predictive of actual deviant behavior several years later. In another classic study, families with children at risk for deviant behavior were randomly assigned to a number of intervention treatments that involved working with the parents or their children in groups. Teaching the parents how to reinforce good behavior in their children was somewhat successful, but meeting with the children in groups perversely *increased* the level of deviant behavior measured a year later. You guessed it: The kids positively reinforced each other for deviancy, which outweighed the instruction that the adults were trying to provide. Any intervention program that involves bringing deviant individuals together in groups—and there are thousands of them— is liable to make the problem worse.

Other examples of prosociality at one level of a multitier social hierarchy becoming part of the problem at higher levels can be recited almost without end. A classic ethnography of a large corporation titled *Moral Mazes* by sociologist Robert Jackall details how competition among individuals, alliances, and divisions within the corporation undermines the goals of the corporation as a whole.[11] Even if a large corporation can manage to function well as a collective unit, there is no guarantee that it will contribute to the welfare of the larger economic system of which it is a part.

A novel titled *No Longer at Ease* by the African author Chinua Achebe sensitively analyzes the phenomenon of corruption through the eyes of an idealistic young man in government service who is full of patriotic fervor for his young nation of Nigeria.[12] Nevertheless, he owes his position to his village, which scrimped and saved to send him to England for his education. As far as his village is concerned, he is supposed to benefit them, regardless of other villages or the nation as a whole. When he becomes guilty of corruption, it is not because he is individually selfish, but because he is obeying the moral dictates of a lower-level social unit. Some nations have managed to solve these internal problems better than others, thereby functioning better as corporate units—but their leaders talk unabashedly about pursuing the national interest, as if blissfully unaware of the dysfunctions that nation-level selfishness can produce at the level of the global village.

In chapter 7 we encountered the so-called first fundamental theorem of welfare economics: "Laissez-faire leads to the common good." The idea that anything like a mathematical proof exists for this proposition, outside the fantasy worlds of neoclassical economics and market fundamentalism, would be funny if it weren't taken so seriously. On the other hand, multilevel selection does have the generality that warrants a term such as theorem or law. In some respects, the law is cause for despair, because it reveals the need for coordination and protection against exploitation at scales that have never existed in human history. In other respects, the law is cause for optimism, because it provides the conceptual resources for achieving functional organization at the planetary scale.

## CHAPTER 10

## *Planetary Altruism*

ALTRUISM EXISTS. If by altruism we mean traits that evolve by virtue of benefitting whole groups, despite being selectively disadvantageous within groups, then altruism indubitably exists and accounts for the group-level functional organization that we see in nature.

Altruism also exists as a criterion that people use for adopting behaviors and policies, with the welfare of whole groups in mind rather than more narrow individual and factional interests. This kind of intentional group selection is as important as natural group selection in the evolution of functionally organized human groups.

Finally, if by altruism we mean a broad family of motives that cause people to score high on a PROSOCIALITY scale by agreeing with statements such as "I think it's important to help other people," then altruism also exists, although more in some people than others. Thanks to the High-PROs of the world, our families, neighborhoods, schools, voluntary associations, businesses, and governments work as well as they do.

Yet this book has been critical of some of the ways that altruism is traditionally studied. Altruism is often defined as a particular psychological motive that leads to other-oriented behaviors, which needs to be distinguished from other kinds of motives. Once the existence of altruism hinges on distinctions among motives, it becomes difficult to study because motives are less transparent than actions.

There are good practical reasons to be concerned about someone's motives, since we want to know how they will behave in the future and in other contexts, not just at the moment. But to the degree that different psychological motives result in the same actions, we shouldn't care much about distinguishing among them, any more than we should care about being paid with cash or a check. It's not right to privilege altruism as a psychological motive when other equivalent motives exist.

In addition, many discussions of altruism implicitly assume a simple relationship between motives and action: that if only everyone *wanted* the world to be better place, then it *would* be. This assumption ignores the complexity and diversity of the proximate mechanisms required for a human social group to function as an organized unit. If we could automatically turn everyone into pure altruists, human society might well collapse despite their best intentions. Why privilege altruism as a psychological motive when so much more is required to understand and accomplish what altruists are motivated to want?

The origin of the word *altruism* is significant in this regard. Recall from chapter 5 that it was coined by Auguste Comte as part of his effort to create a moral system that does not require a belief in God. Comte disagreed with the Catholic doctrine of original sin, which held that people are born entirely sinful and can only find grace by seeking God. Comte maintained that people are naturally capable of both good and evil and that the "great problem of human life" was to organize society so that it functions well without being disrupted by selfish interests. Comte was also striving to capture the moral high ground by centering his religion of humanity on altruism, since valuing the welfare of others for its own sake seems morally superior to valuing the welfare of others as a means to personal salvation. Never mind that Comte's new religion collapsed almost immediately while religions founded on personal salvation are still alive and well. Judged by the actions that they motivate, altruism defined in terms of thoughts and feelings might not be morally superior after all.

Viewed from a twenty-first-century perspective, Comte's aspirations have both a modern and archaic ring. The great problem of human life is still organizing society so that it functions at a larger scale. If by egoism we mean functional organization at smaller scales, then egoism is indeed part of the problem that must be overcome. Science is required to provide solutions, as Comte stressed, including knowledge of animal behavior and the human mind. Scientific knowledge of religious groups as functionally organized units is also helpful. I even admire the panoramic scope of Comte's vision, compared to the shrunken expectations of many scientists and intellectuals today. But Comte's pre-Darwinian understanding of nature, the human mind, and religion are laughably old-fashioned compared to what we know today. Realizing that our own knowledge is still provisional, perhaps we are in a position to succeed where Comte and his contemporaries failed.

The relaunch of Comte's effort needs to be centered on the concept of *functional organization*, not altruism per se. Understanding how groups become functionally organized is a prerequisite for making the world a better place. The fact that single organisms are societies and functionally integrated societies qualify as organisms—not just figuratively but literally—was one of the most important developments of evolutionary thought during the twentieth century. The degree of regulation required for an organism to survive and reproduce in its environment is mind-boggling when studied in detail. Hundreds of metabolic processes must be regulated, such as $CO_2$ levels in the blood, temperature regulation, the hormonal response to a threat, and the sleep cycle. Monitoring and feedback processes are required to forage efficiently, to avoid predators, to ward off disease, to select the best mates, and dozens of other adaptations. For every regulatory mechanism that works, there are countless mechanisms that don't work. All of this order is winnowed, like needles from a haystack of disorder, by the process of natural selection.

Appreciating the complexity of biological organisms helps us appreciate the complexity of small-scale human societies, which evolved by the same process. Natural selection endowed our ancestors with the ability to function as superorganisms, surviving and reproducing by collective action rather than at the expense of each other. Such function required the suppression of disruptive self-serving behaviors within groups and just the right coordinated response to dozens of environmental challenges, which varied across time and geographical location. The mental and physical toolkits required for Inuits to survive their Arctic environment are vastly different from what Bedouins need to survive their desert environment. Genetic evolution endowed us with the mechanisms for deriving and perpetuating these toolkits, including a capacity for symbolic thought that became an inheritance mechanism in its own right. There is no way for this functional and mechanistic complexity to be understood in terms of altruism or selfishness, which is why both concepts must yield center stage to the concept of *organism*.

The invention of agriculture made it possible for human groups to become larger, but the actual historical process was one of multilevel cultural evolution. Larger groups created new problems for coordination and the suppression of disruptive self-serving behaviors. Some groups solved these problems better than others, and the more successful cultural practices spread—by warfare, economic superiority, and imitation. Sometimes the successful practices originated by happenstance, sometimes by intention, but either way, the crucible for their selection was how they contributed to group-level functional organization at a larger scale. Historians have described a rich fossil record of cultural change, but it is only beginning to be interpreted explicitly from a multilevel evolutionary perspective.

One implication of viewing human history as a process of multilevel cultural evolution is that the mechanisms of group-level functional organization in humans are inherently diverse. A common assumption is that human nature can be reduced to a list of psychological universals that result in cultural universals, such as incest avoidance, as if cultural variation is mere noise.[1] This method is not how evolutionary biologists study the diversity of life. Their focus is on how species adapt to their respective environments, even when they share the same genetic inheritance system. By analogy, humans do share a single genetic heritage, including psychological universals, but insofar as these constitute an inheritance system, they result in cultural diversity, not uniformity. We therefore need to focus on how human cultures adapt to their respective environments. Diversity must be the center of attention, not something to be ignored. The question of how human groups become functionally organized as far as proximate causation is concerned probably has no single answer. Different groups do it in different ways, which empirical inquiry must determine. This is another reason to avoid privileging any particular set of proximate mechanisms over another.

Now we are at a point in history when the great problem of human life is to accomplish functional organization at a larger scale than ever. The selection of best practices must be intentional, because we cannot wait for natural selection and there is no process of between-planet selection to select for functional organization at the planetary scale. The problems are daunting but success is still possible when we use evolutionary theory as our navigational guide.

One key insight is that the design principles required for groups to function well are scale-independent. At all scales, there must be mechanisms that coordinate the right kinds of action and prevent disruptive forms of self-serving behavior at lower levels of social organization. The challenge is how to implement these mechanisms at ever-larger scales. This might seem obvious, but only from a certain perspective. As we have seen, those who adopt a "greed is good" perspective believe that we should remove all restrictions on lower-level self-interest. Regulation becomes a dirty word, which is why it is crucial to resist worldviews that depart from factual reality, whether religious or secular, and adopt a perspective based on the best of our scientific knowledge—which means one that is rooted in evolutionary theory.

Regulatory mechanisms do not necessarily require centralized planning or top-down regulation. Indeed, this kind of regulation is very likely to fail. Human social systems are so complex that no one knows how they work, and even the best-formulated plans are likely to fail due to unforeseen consequences. The solution to this problem is to experiment with new social arrangements, monitor naturally occurring variation, and cautiously adopt what works. In other words, we need to become wise managers of variation and selection processes.[2]

The need to manage self-organizing processes might seem like a contradiction in terms, but it follows directly from evolutionary theory. Most regulatory processes in biological organisms are self-organizing, but a selection process was required to winnow the self-organizing processes that work from the many that don't work. We must play the role of selective agent for self-organizing processes in modern life. Adaptive self-organization at large societal scales will not emerge as a robust property of self-interest or complex systems in the absence of selection. This conclusion follows from evolutionary theory at such a basic level that it is unlikely to be wrong.

Part of the import of Elinor Ostrom's work is to highlight the importance of small groups as units of functional organization. They are often best qualified to regulate themselves and adapt to their local environments. Small groups require the core design principles, as do groups of any size, but the core design principles are often more easily realized in small groups than in larger groups. From an evolutionary perspective, we can say that large-scale human society needs to be multicellular. The more we participate in small groups that are appropriately structured, the happier we will be, the more our group efforts will succeed, and the more we will contribute to the welfare of society at larger scales. With this in mind, the Evolution Institute is creating a practical framework for improving the efficacy of groups called PROSOCIAL, which makes it easy for groups to learn about and adopt the core design principles to the extent they have not spontaneously done so.[3] It's good to know that we can begin to make the world a better place by strengthening the groups that are already most salient to our everyday lives.

Ostrom also pioneered the concept of *polycentric governance*, which notes that human life consists of many spheres of activity and that each sphere has an optimal scale.[4] Optimal governance requires determining the optimal scale for each sphere of activity and coordinating appropriately among spheres. Stated this way it can scarcely seem otherwise, but most large-scale governance does not operate this way. Thinking of large-scale society as a multicellular organism can help to generalize and refine the concept of polycentric governance, as it has the governance that takes place in small groups.

Most of the ideas that I have reported in this book are recent, especially when compared to the history of thinking on altruism. The concept of major evolutionary transitions, which merges the concepts of organism and society, wasn't proposed until the 1970s. Multilevel selection theory's road to reacceptance was so long that this is one of the first books to offer a postresolution account. The use of evolution to justify social inequality and ruthless competition stigmatized the study of evolution in relation to human affairs for decades following World War II. A renewed effort didn't gather steam until the late twentieth century. The first book-length treatments of religion from a modern evolutionary perspective didn't appear until the start of the twenty-first century. Rethinking economics and public policy from an evolutionary perspective is more recent still. For the vast majority of politicians and policymakers today, evolution is a word to be avoided. If it stands for anything, it is the cruel policies that became associated with the term *social Darwinism* in the first half of the twentieth century.

These recent ideas are as foundational as the ideas associated with the Enlightenment and the early days of evolutionary theory. In this book, I have endeavored to show how they bear upon the topic of altruism. Some of my conclusions are provisional, but others follow from evolutionary theory at such a basic level that they are unlikely to be wrong. The distinction between proximate and ultimate causation should forever change the way that altruism defined at the level of thoughts and feelings is conceptualized. And multilevel selection theory makes it crystal clear that if we want the world to become a better place, we must choose policies with the welfare of the whole world in mind. As far as our selection criteria are concerned, we must become planetary altruists.

# Notes

**INTRODUCTION**

1. Neusner and Chilton (2005).
2. Dugatkin (2006).
3. My previous book on altruism with Elliott Sober (Sober and Wilson 1998) provides more depth, especially with respect to psychological altruism.
4. I recount these stories in detail in Wilson (2011).

**CHAPTER 1**

1. Hutchins (1996).
2. Ostrom (1990, 2010).
3. Hardin (1968).
4. McGinnis (1999).
5. Cox, Arnold, and Villamayor-Thomas. (2010).
6. Wilson, Ostrom, and Cox (2013).
7. Holldobler and Wilson (2008).
8. Seeley (1995, 2010); Seeley and Buhrman (1999); Seeley et al. (2012); Passino, Seely, and Visscher. (2007).
9. Described in Seeley et al. (2012).
10. Prins (1995).
11. Sontag, Wilson, and Wilcox (2006).
12. Couzin (2007); Couzin et al. (2011); Wilson (2000).

**CHAPTER 2**

1. See especially Frank (2011).
2. Wilson and Wilson (2007, p. 348).

3. Chapters 3 and 4 of *The Neighborhood Project* (Wilson 2011) are titled "The Parable of the Strider" and "The Parable of the Wasp." They are written to illustrate the difference between within- and between-group selection.
4. Eldakar et al. (2009a).
5. Eldakar et al. (2009b; 2010).
6. Kerr et al. (2006).
7. Margulis (1970).
8. Maynard Smith and Szathmary (1995, 1999); see Bourke (2011) and Calcott and Sterelny (2011) for recent treatments.
9. Burt and Trivers (2006); Crespi and Summers (2005); Pepper et al. (2009).

## CHAPTER 3

1. Darwin (1871, p. 166).
2. See Sober and Wilson (1998); Borrello (2010); Harmon (2010); and Okasha (2006) for historical and conceptual accounts.
3. Ghiselin (1974, p. 247); Alexander (1987, p. 3).
4. Hamilton (1975) is the seminal paper in this regard. Hamilton initiated inclusive fitness theory in the 1960s and regarded it as an alternative to group selection at the time. After encountering the work of George Price, he reformulated his own theory and realized that it invoked selection within and among groups after all, which he stated clearly in his 1975 paper. The fact that the entire community of evolutionary biologists did not quickly follow suit is a puzzle for social historians to solve. Hamilton (1996) describes his "conversion" in his own words, and Harmon (2010) has written an excellent book on Price and Hamilton.
5. My first article on group selection appeared in 1975, the same year that Hamilton's article was published, and coined the term "trait group" to define the sets of individuals within which social interactions occur for any given evolving trait. I provided a simple algebraic model showing that selection among trait groups in a multigroup population can easily counteract selection among individuals within trait groups, contrary to the received wisdom about group selection. No one could dispute the math, but the typical response to the paper was "Why would you call that 'group selection'?"
6. See Dugatkin and Reeve (1994); Kerr and Godfrey-Smith (2002); Marshall (2011); Queller (1991), Okasha (2006, 2014, Wilson (2008), Wilson and Sober (2002). A summary of Okasha's 2006 book *Evolution and the Levels of Selection* and commentaries followed by a reply was published in the journal *Philosophy and Phenomenological Research* (Okasha 2011a, 2011b; Sober 2011; Waters 2011).
7. Popper (1934).
8. Kuhn (1970).

9. Campbell and Grondona (2010).
10. Gintis (2009).
11. Before leaving the topic of equivalence, it is important to stress that evolutionary theories of social behavior can be equivalent in some respects without being equivalent in all respects. In particular, accounting methods that average the fitness of individuals across groups are straightforward in additive models but become cumbersome and perhaps even impossible with nonadditive effects. An example is the concept of equilibrium selection, in which complex social interactions result in multiple local equilibria that differ in their group-level properties. Group selection favors local equilibria that have the highest group fitness, but the component traits are also positively selected within groups. It is hard to see how this scenario can be modeled by inclusive fitness theory. For a sample of the recent literature on this topic, see Traulsen (2010); Marshall (2011); van Veelen et al. (2012); Goodnight (2013); Simon, Fletcher, Doebel (2013); and Simon (2014).

## CHAPTER 4

1. Once a major transition occurs, many species can result from the first species. Thus, the origination of the first eusocial wasps, bees, ants, and termites was rare, not their subsequent adaptive radiation.
2. Haig (1997); Burt and Trivers (2006); Pepper et al. (2009).
3. See Burkart and van Schaik (2012) and de Waal (2009) for a review of cooperation in nonhuman primates.
4. Boehm (1993, 1999, 2012); Bingham (1999); Bingham and Souza (2011).
5. Henrich and Gil-White (2001).
6. E. O. Wilson (2012).
7. Wilson (2002).
8. Wegner (1986, p.185).
9. Campbell (1994, p. 23). Donald T. Campbell (1916–1996) was a social psychologist who pioneered the study of altruism, group-level functional organization, and cultural evolution from a modern evolutionary perspective.
10. Wilson (2009); see also Bingham (1999); Bingham and Souza (2011); Boehm (1999, 2012); Tomasello (2009).
11. Tomasello et al. (2005); Tomasello (2009).
12. Hare et al. (2002); Hare and Woods (2013).
13. Deacon (1998); Jablonka and Lamb (2005).
14. Wilson, Ostrom, and Cox. (2013).
15. Jablonka and Lamb (2005). Calling symbolic thought a nongenetic inheritance system does not deny that it evolved by genetic evolution. The point is that

after it evolved, it creates heritable phenotypic variation that is independent of genetic variation.

16. Jablonka and Lamb (2005) provide a concise history of how evolutionary biology became so gene-centric.
17. See, for example, Smith (2003).
18. Gissis and Jablonka (2011).
19. Gregory and Webster (1996).
20. West-Eberhard (2003); Piersma and van Gils (2010).
21. E. O. Wilson (2012).
22. Pagel and Mace (2004); Pagel (2012).
23. The concept of cultural traits as like parasites and disease organisms has been developed by Dawkins (1976); Blackmore (1999); Dennett (2006).
24. Richerson and Boyd (2005).
25. Richerson and Boyd (1999); Stoelhorst and Richerson (2013); Turchin (2005, 2011).
26. Turchin (2005); Acemoglu and Robinson (2012).

## CHAPTER 5

1. Sober (1984).
2. Williams (1966); see Sober and Wilson (1998) for a more detailed treatment.
3. Colwell (1981); Wilson and Colwell (1981).
4. See Sober and Wilson (1998) for a more detailed account.
5. Wilson and Sober (1989, 1994); Sober and Wilson (1998).
6. Batson (1991, 2011).
7. Mayr (1961).
8. Tinbergen (1963); see Laland et al. (2011) and Scott-Phillips , Dickens, and West. (2011) for recent treatments.
9. Holden and Mace (2009).
10. This thought experiment is similar to one that Elliott Sober and I conducted to compare hedonists, egoists, and altruists (Sober and Wilson 1998), but these distinctions do not exactly correspond to the three types of individuals discussed here.
11. The veil of ignorance made famous by John Rawls in his *Theory of Justice* (1971) assumes that the average person reasons like Dick. The veil is required for Dick to choose social arrangements that qualify as just.
12. These nuances are based on mathematical models in which evolution takes place in a single group (the desert island scenario) or in multiple groups with random variation among groups (Wilson 1975, 1977).
13. See also Haidt (2012).

## CHAPTER 6

1. Boehm (2012).
2. Ehrenpreis, Felbringer, and Friedmann. (1978, p. 11).
3. Wilson (2002).
4. Atran (2002); Boyer (2001).
5. Dawkins (2006); Dennett (2006); Hitchens (2007); Harris (2004).
6. Dennett (1995).
7. Tyler (1871); Frazer (1890); Durkheim and Fields (1912).
8. Gould and Lewontin (1979).
9. In his introduction to the second edition of his book on Darwin's finches, Lack (1961) notes the change in thinking that had occurred since the publication of the first edition. My article on adaptive individual differences within single populations (Wilson 1998) notes the trend during the subsequent decades showing that natural selection operates on much finer spatial and temporal scales than Lack knew.
10. Wright (2009); Wade (2009); Norenzayan (2013); Bellah (2011).
11. One of my contributions to this consensus is an analysis of a random sample of religions, which avoids selection bias (Wilson 2005).
12. Ehrenpreis, Felbringer, and Friedmann. (1978); discussed in more detail by Wilson and Sober (1994).
13. Neusner and Chilton (2005); also see Neusner and Chilton (2009) for a similarly organized volume on the Golden Rule in world religions.
14. Berchman (2005, p. 2).
15. Wilson (1995).
16. Neusner and Chilton (2005, p. 191).
17. Numbers (1976); for a more detailed account of this example, see Wilson (2011, ch. 18).
18. Hafen and Hafen (1992).
19. In my study of a random sample of religions (Wilson 2005), the frequency of religions that spread by violent conflict was very low.
20. Comte (1851).
21. Dixon (2005, p. 203).
22. Dixon (2005, p. 205).

## CHAPTER 7

1. Wilson (2007).
2. Like Gandolf, Bernard rejoined the Evolution Institute and played an integral role in its development until his death in 2014. He will be missed.

3. The special issue at the *Journal for Economic Behavior and Organization* website, http://www.sciencedirect.com/science/journal/01672681/90/supp/S. Short accessible articles describing each of the thirteen articles in the special issue are available at the online evolution magazine *This View of Life*, http://www.thisviewoflife.com/index.php/magazine/articles/evolution-and-economics-special-issue.

4. See Wilson (2011, ch. 19) for a narrative account of my journey.

5. Smith (1759, 1776).

6. Burgin (2012); Jones (2012).

7. Rand (1957).

8. Mandeville (1705).

9. Wright (2005, p. 46).

10. Beinhocker (2005, p. 30).

11. Veblen (1898, p. 389).

12. Thaler and Sunstein (2008, p. 6).

13. Friedman (1953).

14. Hodgson (1991); Stone (2010); Lewis (2012).

15. Gould and Lewontin (1979).

16. Feldman (2008) discussed from an evolutionary perspective by Wilson (2012a).

17. http://www.nydailynews.com/news/politics/idaho-legislator-suggests-making-ayn-rand-required-high-school-reading-article-1.1257262.

18. http://www.aynrand.org/campus.

19. Wilson (1995).

20. Rand (1961, p. 50).

21. http://www.theawl.com/2011/04/when-alan-met-ayn-atlas-shrugged-and-our-tanked-economy.

22. Branden (1989, p.7).

23. http://www.nytimes.com/2008/10/24/business/economy/24panel.html?_r=0.

24. Wilson and Gowdy (2015).

25. Gintis et al. (2005).

26. Tocqueville (1835, p. 60).

27. Pickett and Wilkinson (2009); Acemoglu and Robinson (2012); Piketty (2014).

28. Lewis (2014).

29. Lewis (2014, p. 164).

30. Strobel et al. (2011).

31. Lewis (2014, p. 95).

## CHAPTER 8

1. Wilson (2011) provides a book-length description of the Binghamton Neighborhood Project and the evolutionary paradigm that informs it for a general audience. Wilson et al. (2014) and Wilson and Gowdy (2013) provide overviews of how evolutionary theory provides a general theoretical framework for positive change at scales ranging from single individuals to small groups to large populations.
2. Visit the Evolution Institute website (http://evolution-institute.org) for more on these projects.
3. See http://www.search-institute.org.
4. Wilson, O'Brien, and Sesma. (2009).
5. Physical traits can be flexible in the same way. The general term that evolutionists use for traits that vary in response to the environment during the lifetime of the organism is *phenotypic plasticity*. A large literature exists on phenotypic plasticity that is highly relevant to understanding and improving the human condition. See Pigliucci (2001); West-Eberhard (2003); DeWitt and Scheiner (2004); and Piersma and van Gils (2010) for book-length academic treatments.
6. These results are reported in Wilson (2011); Wilson, O'Brien, and Sesma. (2009); Wilson and O'Brien (2010); O'Brien (2012); O'Brien, Gallup, and Wilson (2012); O'Brien and Kauffman (2012); and O'Brien, Norton, Cohen, and Wilson (2012).
7. Prosociality from an evolutionary perspective intersects with concepts such as social capital and collective efficacy developed by sociologists such as Robert Putnam (2000) and Robert Sampson (2003, 2004).
8. Bjorklund (2007); Del Giudice, Ellis, and Shirtcliff. (2011); Ellis and Bjorklund (2005); Ellis et al. (1999, 2011).
9. Gluckman and Hanson (2004); Bateson and Gluckman (2011).
10. Carey (2011); Jablonka and Lamb (2005).
11. Ellis et al. (2011); Del Giudice, Ellis, and Shirtcliff. (2011). Science writer David Dobbs has reported on this subject in the December 2009 issue of *Atlantic Monthly* and elsewhere; see http://www.theatlantic.com/magazine/archive/2009/12/the-science-of-success/307761/.
12. O'Brien and Wilson (2011).
13. O'Brien, Gallup and Wilson (2012).
14. Wilson, Ostrom, and Cox (2013).
15. Csikszentmihalyi and Schneider (2000).
16. Wilson, Kauffman, and Purdy (2011).
17. Kauffman and Wilson (2014).
18. Embry (2002); Bradshaw et al. (2009); Kellam et al. (2008).

19. Prinz et al. (2009). Visit the Triple P website (http://www.triplep.net/glo-en/home/) to learn more about this remarkable program.
20. Wilson et al. (2014).
21. I have accumulated considerable experience on this point. In my communitarian projects, I frequently work with religious believers of all stripes. You would think that being a professed atheist and evolutionist would be a disadvantage, but I am typically judged primarily for the skills that I bring to the table, my commitment to a common cause, and my tolerance of their meaning systems.
22. Grant (2013). An interview that I conducted with Grant for the online evolution magazine *This View of Life* is available at http://www.thisviewoflife.com/index.php/magazine/articles/prosociality-works-in-the-workplace.

### CHAPTER 9

1. Oakley et al. (2011).
2. Ellis et al. (2012).
3. Wilson and Csikszentmihalyi (2007).
4. O'Conner et al. (2011).
5. McGrath and Oakley (2011, ch. 4); Widiger and Presnall (2011, ch. 6); Zahn-Waxler and Van Hulle (2011, ch. 25).
6. Bachner-Melman (2011, ch. 7).
7. Li and Rodin (2011, ch. 11); Klimecki and Singer (2012, ch. 28).
8. Oakley (2011).
9. Exline et al. (1970).
10. Capaldi et al. (2001).
11. Jackall (2009).
12. Achebe (1960).

### CHAPTER 10

1. Brown (1991).
2. Wilson et al. (2014).
3. View this video (http://alanhonick.com/prosocial) and visit the website of the Evolution Institute (http://evolution-institute.org/) to learn more.
4. This concept was also pioneered by Elinor Ostrom's husband, Vincent Ostrom. See McGinnis (1999) for a volume of collected papers.

# Works Cited

Acemoglu, D., & Robinson, J. (2012). *Why Nations Fail: The Origins of Power, Prosperity, and Poverty.* New York: Crown.

Achebe, C. (1960). *No Longer at Ease.* New York: Anchor.

Alexander, R. D. (1987). *The Biology of Moral Systems.* New York: Aldine de Gruyter.

Atran, S. (2002). *In Gods We Trust: The Evolutionary Landscape of Religion.* Oxford: Oxford University Press.

Atran, S., & Henrich, J. (2010). "The Evolution of Religion: How Cognitive By-Products, Adaptive Learning Heuristics, Ritual Displays, and Group Competition Generate Deep Commitments to Prosocial Religions." *Biological Theory: Integrating Development, Evolution, and Cognition, 5,* 18–30.

Bachner-Melman, R. (2011). The Relevance of Pathological Altruism to Eating Disorders. In B. Oakley, A. Knafo, G. Madhavan, & D. S. Wilson (Eds.), *Pathological Altruism* (pp. 94–106). Oxford: Oxford University Press.

Bateson, P., & Gluckman, P. (2011). *Plasticity, Robustness, Development, and Evolution.* New York: Cambridge University Press.

Batson, C. D. (1991). *The Altruism Question: Toward a Social-Psychological Answer.* Hillsdale, NJ: Lawrence Erlbaum Associates.

Batson, C. D. (2011). *Altruism in Humans.* First ed. Oxford: Oxford University Press.

Bellah, R. (2011). *Religion in Human Evolution: From the Paleolithic to the Axial Age.* Cambridge, MA: Belknap Press.

Berchman, R. M. (2005). Altruism in Greco-Roman Philosophy. In J. Neusner & B. Chilton (Eds.), *Altruism in World Religions* (pp. 1–30). Washington, D.C.: Georgetown University Press.

Bingham, P. M. (1999). "Human Uniqueness: A General Theory." *Quarterly Review of Biology, 74,* 133–169.

Bingham, P. M., & Souza, J. (2011). *Death from a Distance and the Birth of a Humane Universe.* BookSurge.

Bjorklund, D. F. (2007). *Why Youth Is Not Wasted on the Young: Immaturity in Human Development.* New York: Wiley-Blackwell.

Blackmore, S. (1999). *The Meme Machine.* Oxford: Oxford University Press.

Boehm, C. (1993). "Egalitarian Society and Reverse Dominance Hierarchy." *Current Anthropology, 34,* 227–254.

Boehm, C. (1999). *Hierarchy in the Forest.* Cambridge, MA: Harvard University Press.

Boehm, C. (2012). *Moral Origins: The Evolution of Virtue, Altruism, and Shame.* New York: Basic Books.

Borrello, M. (2010). *Evolutionary Restraints: The Contentious History of Group Selection.* Chicago: University of Chicago Press.

Bourke, A. F. G. (2011). *Principles of Social Evolution.* Oxford: Oxford University Press.

Boyer, P. (2001). *Religion Explained.* New York: Basic Books.

Bradshaw, C. P., Zmuda, J. H., Kellam, S., & Ialongo, N. (2009). "Longitudinal Impact of Two Universal Preventive Interventions in First Grade on Educational Outcomes in High School." *Journal of Educational Psychology, 101,* 926–937.

Branden, N. (1989). *Judgment Day.* Boston: Houghton Mifflin.

Brown, D. (1991). *Human Universals.* First ed. New York: McGraw-Hill Humanities / Social Sciences / Languages.

Burgin, A. (2012). *The Great Persuasion: Reinventing Free Markets since the Depression.* Cambridge, MA: Harvard University Press.

Burkart, J. M., & van Schaik, C. (2012). "Group Service in Macaques (*Macaca Fuscata*), Capuchins (*Cebus Apella*), and Marmosets (*Callithrix Jacchus*): A Comparative Approach to Identifying Proactive Prosocial Motivations." *Journal of Comparative Psychology.* doi:10.1037/a0026392

Burt, A., & Trivers, R. (2006). *Genes in Conflict.* Cambridge, MA: Harvard University Press.

Calcott, B., & Sterelny, K. (Eds.). (2011). *The Major Transitions in Evolution Revisited.* Cambridge, MA: MIT Press.

Campbell, D. T. (1994). How Individual and Face-to-Face Group Selection Undermine Firm Selection in Organizational Evolution. In J. A. C. Baum & J. V. Singh (Eds.), *Evolutionary Dynamics of Organizations* (pp. 23–38). New York: Oxford University Press.

Campbell, L., & Grondona, V. (2010). "Who Speaks What to Whom? Multilingualism and Language Choice in Misión La Paz." *Language in Society, 39*(5), 617–646. doi:10.1017/S0047404510000631

Capaldi, D. M., Dishion, T. J., Stoolmiller, M., & Yoerger, K. (2001). "Aggression toward Female Partners by At-Risk Young Men: The Contribution of Male Adolescent Friendships." *Developmental Psychology, 37*, 61–73.

Carey, N. (2011). *The Epigenetics Revolution: How Modern Biology Is Rewriting Our Understanding of Genetics, Disease, and Inheritance.* New York: Columbia University Press.

Colwell, R. K. (1981). "Group Selection Is Implicated in the Evolution of Female-Biased Sex Ratios." *Nature, 290*, 401–404.

Comte, A. (1851). *Systeme de Politique Positive.* (Paris: n.p.)

Couzin, I. (2007). "Collective Minds." *Nature, 445*(7129), 715. doi:10.1038/445715a

Couzin, I. D., Ioannou, C. C., Demirel, G., Gross, T., Torney, C., Hartnett, A., . . . Leonard, N. E. (2011). "Democratic Consensus in Animal Groups." *Science, 334*, 1578–1580

Cox, M., Arnold, G., & Villamayor-Tomas, S. (2010). "A Review of Design Principles for Community-Based Natural Resource Management." *Ecology and Society, 15(4).* http://www.ecologyandsociety.org/vol15/iss4/art38/

Crespi, B., & Summers, K. (2005). "Evolutionary Biology of Cancer." *Trends in Ecology & Evolution, 20*(10), 545–552. doi:10.1016/j.tree.2005.07.007.

Csikszentmihalyi, M., & Schneider, B. (2000). *Becoming Adult: How Teenagers Prepare for the World of Work.* New York: Basic Books.

Darwin, C. (1871). *The Descent of Man and Selection in Relation to Sex.* London: John Murray.

Dawkins, R. (1976). *The Selfish Gene.* First ed. Oxford: Oxford University Press.

Dawkins, R. (2006). *The God Delusion.* Boston: Houghton Mifflin.

Deacon, T. W. (1998). *The Symbolic Species.* New York: Norton.

Del Giudice, M., Ellis, B. J., & Shirtcliff, E. A. (2011). "The Adaptive Calibration Model of Stress Responsivity." *Neuroscience and Biobehavioral Reviews, 35*(7), 1562–1592. doi:10.1016/j.neubiorev.2010.11.007

Dennett, D. C. (1995). *Darwin's Dangerous Idea: Evolution and the Meanings of Life.* New York: Simon and Schuster.

Dennett, D. C. (2006). *Breaking the Spell: Religion as a Natural Phenomenon.* New York: Viking.

de Waal, F. (2009). *The Age of Empathy: Nature's Lessons for a Kinder Society.* New York: Crown.

DeWitt, T. J., & Scheiner, S. M. (Eds.). (2004). *Phenotypic Plasticity: Functional and Conceptual Approaches.* Oxford: Oxford University Press.

Dixon, T. (2005). The Invention of Altruism: Auguste Comte's Positive Polity and Respectable Unbelief in Victorian Britain. In D. M. Knight & M. D. Eddy (Eds.), *Science and Beliefs: From Natural Philosophy to Natural Science, 1700–1900* (pp. 195–212). Burlington, VT: Ashgate.

Dugatkin, L. A. (2006). *The Altruism Equation: Seven Scientists in Search for the Origins of Goodness.* Princeton, NJ: Princeton University Press.

Dugatkin, L. A., & Reeve, H. K. (1994). "Behavioral Ecology and Levels of Selection: Dissolving the Group Selection Controversy." *Advances in the Study of Behavior, 23*, 101–133.

Durkheim, E., & Fields, K. E. (1912). *The Elementary Forms of Religious Life.* New York: Free Press.

Ehrenpreis, A., Felbinger, C., & Friedmann, R. (1978). An Epistle on Brotherly Community as the Highest Command of Love. In R. Friedmann (Ed.), *Brotherly Community: The Highest Command of Love* (pp. 9–77). Rifton, NY: Plough Publishing.

Eldakar, O. T., Dlugos, M., Wilcox, R. S. and Wilson, D. S. (2009a) Aggressive Mating as a Tragedy of the Commons in a Water Strider Aquarius Remigis. *Behavioral Ecology and Sociobiology* (pp. 64, 25–33).

Eldakar, O. T. Dlugos, M. J., Holt, G. P., Wilson, D. S., & Pepper, J. W. (2009b). "Population Structure Mediates Sexual Conflict in Wild Populations of Water Striders." *Science, 326*, 816. doi:10.1126/science.1180183

Eldakar, O. T., Wilson, D. S., Dlugos, M. J., & Pepper, J. W. (2010). "The Role of Multilevel Selection in the Evolution of Sexual Conflict in the Water Strider *Aquarius Remigis.*" *Evolution: International Journal of Organic Evolution, 64*(11), 3183–3189.

Ellis, B. J., & Bjorklund, D. F. (2005). *Origins of the Social Mind: Evolutionary Psychology and Child Development.* New York: Guilford Press.

Ellis, B. J., Boyce, W. T., Belsky, J., Bakermans-Kranenburg, M. J., & van Ijzendoorn, M. H. (2011). "Differential Susceptibility to the Environment: An Evolutionary–Neurodevelopmental Theory." *Development and Psychopathology, 23*(1), 7–28. doi:10.1017/S0954579410000611

Ellis, B. J., Del Giudice, M., Dishion, T. J., Figueredo, A. J., Gray, P., Griskevicius, V., . . . Wilson, D. S. (2012). "The Evolutionary Basis of Risky Adolescent Behavior: Implications for Science, Policy, and Practice." *Developmental Psychology, 48*(3), 598–623. doi:10.1037/a0026220

Ellis, B. J., McFadyen-Ketchum, S., Dodge, K. A., Pettit, G. S., & Bates, J. E. (1999). "Quality of Early Family Relationships and Individual Differences in the Timing of Pubertal Maturation in Girls: A Longitudinal Test of an Evolutionary Model." *Journal of Personality and Social Psychology, 77*, 387–401.

Embry, D. D. (2002). "The Good Behavior Game: A Best Practice Candidate as a Universal Behavioral Vaccine." *Clinical Child and Family Psychology Review, 5*, 273–297.

Exline, R. V., Thiabaut, J., Hickey, C. B., Bumpart, P., Christie, R., & Geis, F. (1970). *Studies in Machiavellianism.* New York: Academic Press.

Feldman, A. M. (2008). Welfare Economics. In S. N. Durlauf & L. E. Blume, Eds.) *The New Palgrave Dictionary of Economics.* New York: Palgrave Macmillan.

Frank, R. (2011). *The Darwin Economy: Liberty, Competition, and the Common Good.* Princeton, NJ: Princeton University Press.

Frazer, J. G. (1890). *The Golden Bough.* London: MacMillan.

Friedman, M. (1953). *Essays in Positive Economics.* Chicago: University of Chicago Press.

Ghiselin, M. T. (1974). *The Economy of Nature and the Evolution of Sex.* Berkeley: University of California Press.

Gintis, H. (2009). *Game Theory Evolving* (Second ed.). Princeton, NJ: Princeton University Press.

Gintis, H., Bowles, S., Boyd, R., & Fehr, E. (2005). *Moral Sentiments and Material Interests.* The Foundations of Cooperation in Economic Life. Cambridge, MA: MIT Press.

Gissis, S. B., & Jablonka, E. (2011). *Transformations of Lamarckism: From Subtle Fluids to Molecular Biology.* Cambridge, MA: MIT Press.

Gluckman, P., & Hanson, M. (2004). *The Fetal Matrix: Evolution, Development, and Disease.* Cambridge: Cambridge University Press.

Goodnight, C. (2013). "On Multilevel Selection and Kin Selection: Contextual Analysis Meets Direct Fitness." *Evolution: International Journal of Organic Evolution, 67*(6), 1539–1548. doi:10.1111/j.1558-5646.2012.01821.x

Gould, S. J., & Lewontin, R. C. (1979). "The Spandrels of San Marco and the Panglossian Paradigm: A Critique of the Adaptationist Program." *Proceedings of the Royal Society of London, B205,* 581–598.

Grant, A. M. (2013). *Give and Take: A Revolutionary Approach to Success.* New York: Viking.

Gregory, S. W. J., & Webster, S. (1996). "A Nonverbal Signal in Voices of Interview Partners Effectively Predicts Communication Accommodation and Social Status Perceptions." *Journal of Personality and Social Psychology, 70,* 1231–1240.

Hafen, L. R., & Hafen, A. W. (1992). *Handcarts to Zion: The Story of a Unique Western Migration, 1856–1860.* Lincoln, NE: Bison Books.

Haidt, J. (2012). *The Righteous Mind: Why Good People Are Divided by Politics and Religion.* New York: Pantheon.

Haig, D. (1997). The Social Gene. In J. R. Krebs & N. B. Davies (Eds.), *Behavioural Ecology: An Evolutionary Approach* (pp. 284–304). Oxford: Blackwell.

Hamilton, W. D. (1975). Innate Social Aptitudes in Man: An Approach from Evolutionary Genetics. In R. Fox (Ed.), *Biosocial Anthropology* (pp. 133–155). London: Malaby Press.

Hamilton, W. D. (1996). *The Narrow Roads of Gene Land.* (Vol. 1). Oxford: W. H. Freeman/Spektrum.

Hardin, G. (1968). "The Tragedy of the Commons." *Science, 162,* 1243–1248.

Hare, B., Brown, M., Williamson, C., & Tomasello, M. (2002). "The Domestication of Social Cognition in Dogs." *Science, 298*, 1634–1636.

Hare, B., & Woods, V. (2013). *The Genius of Dogs: How Dogs Are Smarter Than You Think.* New York: Dutton.

Harmon, O. S. (2010). *The Price of Altruism.* New York: Norton.

Harris, S. (2004). *The End of Faith: Religion, Terror, and the Future of Reason.* New York: Norton.

Henrich, J., & Gil-White, F. J. (2001). "The Evolution of Prestige: Freely Conferred Deference as a Mechanism for Enhancing the Benefits of Cultural Transmission." *Evolution and Human Behavior, 22*(3), 165–196. doi:10.1016/S1090-5138(00)00071-4

Hitchens, C. (2007). *God Is Not Great: How Religion Poisons Everything.* New York: Twelve Books.

Hodgson, G. M. (1991). Hayek's Theory of Cultural Evolution: An Evaluation in the Light of Vanberg's Critique. *Economics and Philosophy, 7*(01), 67–82. doi:10.1017/S0266267100000912

Holden, C., & Mace, M. (2009). "Phylogenetic Analysis of the Evolution of Lactose Digestion in Adults." *Human Biology, 81*, 597–619.

Holldobler, B., & Wilson, E. O. (2008). *The Superorganisms.* New York: Norton.

Hutchins, E. (1996). *Cognition in the Wild.* Boston: MIT Press.

Jablonka, E., & Lamb, M. (2005). *Evolution in Four Dimensions: Genetic, Epigenetic, Behavioral, and Symbolic Variation in the History of Life.* Cambridge, MA: MIT Press.

Jackall, R. (2009). *Moral Mazes: The World of Corporate Managers.* New York: Oxford University Press.

Jones, D. S. (2012). *Masters of the Universe: Hayek, Friedman, and the Birth of Neoliberal Politics.* Princeton, NJ: Princeton University Press.

Kauffman, R. A., & Wilson, D. S. (2014). "Effect of a Program for At-Risk Youth on Non-Academic Developmental Assets." Manuscript in progress.

Kellam, S., Brown, C. H., Poduska, J., Ialongo, N., Wang, W., Toyinbo, P., . . . Wilcox, H. C. (2008). "Effects of a Universal Classroom Behavior Management Program in First and Second Grades on Young Adult Behavioral, Psychiatric, and Social Outcomes." *Drug and Alcohol Dependence, 95*, S5–S28.

Kerr, B., & Godfrey-Smith, P. (2002). "Individualist and Multi-level Perspectives on Selection in Structured Populations." *Biology and Philosophy, 17*(4), 477–517. doi:10.1023/A:1020504900646

Kerr, B., Neuhauser, C., Bohannan, B. J. M., & Dean, A. M. (2006). "Local Migration Promotes Competitive Restraint in a Host-Pathogen 'Tragedy of the commons.'" *Nature, 442*, 75–78.

Klimecki, O., & Singer, T. (2011). Empathetic Distress Fatigue Rather than Compassion Fatigue? Integrating Findings from Empathy Research in Psychology and Social Neuroscience. In B. Oakley, A. Knafo, G. Madhavan, & D. S. Wilson (Eds.), *Pathological Altruism* (pp. 368–384). Oxford: Oxford University Press.

Kuhn, T. S. (1970). *The Structure of Scientific Revolutions.* Second ed. Chicago: University of Chicago Press.

Lack, D. (1961). *Darwin's Finches.* Cambridge: Cambridge University Press.

Laland, K., Sterelny, K., Odling-Smee, F. J., Hoppitt, W., & Uller, T. (2011). "Cause and Effect in Biology Revisited: Is Mayr's Proximate-Ultimate Dichotomy Still Useful?" *Science, 334*, 1512–1516.

Lewis, M. (2014). *Flash Boys: A Wall Street Revolt.* New York: Norton.

Lewis, P. (2012). "Emergent Properties in the Work of Friedrich Hayek." *Journal of Economic Behavior and Organization, 82*(2–3), 368–378. doi:10.1016/j.jebo.2011.04.009

Li, M., & Rodin, G. (2011). Altruism and Suffering in the Context of Cancer: Implications of the Relational Paradigm. In B. Oakley, A. Knafo, G. Madhavan, & D. S. Wilson (Eds.), *Pathological Altruism* (pp. 138–155). Oxford: Oxford University Press.

Mandeville, B. (1705). *The Fable of the Bees: or Private Vices, Publick Benefits.* Indianapolis: Liberty Press.

Margulis, L. (1970). *Origin of Eukaryotic Cells.* New Haven, CT: Yale University Press.

Marshall, J. A. R. (2011). "Group Selection and Kin Selection: Formally Equivalent Approaches." *Trends in Ecology and Evolution, 26*(7), 325–332. doi:10.1016/j.tree.2011.04.008.

Maynard Smith, J., & Szathmary, E. (1995). *The Major Transitions of Life.* New York: W. H. Freeman.

Maynard Smith, J., & Szathmary, E. (1999). *The Origins of Life: From the Birth of Life to the Origin of Language.* Oxford: Oxford University Press.

Mayr, E. (1961). "Cause and Effect in Biology." *Science, 134*(3489), 1501–1506.

McGinnis, M. D. (1999). *Polycentric Governance and Development: Readings from the Workshop in Political Theory and Policy Analysis.* Ann Arbor: University of Michigan Press.

McGrath, M., & Oakley, B. (2011). Codependency and Pathological Altruism. In B. Oakley, A. Knafo, G. Madhavan, & D. S. Wilson (Eds.), *Pathological Altruism* (pp. 49–74). Oxford: Oxford University Press.

Neusner, J., & Avery-Peck, A. J. (2005). Altruism in Classical Judaism. In J. Neusner & B. Chilton (Eds.), *Altruism in World Religions* (pp. 31–52). Washington, D.C.: Georgetown University Press.

Neusner, J., & Chilton, B. (2005). *Altruism in World Religions.* Washington, D.C.: Georgetown University Press.

Neusner, J., & Chilton, B. D. (2009). *The Golden Rule: The Ethics of Reciprocity in World Religions.* New York: Continuum.

Norenzayan, A. (2013). *Big Gods: How Religion Transformed Cooperation and Conflict.* Princeton, NJ: Princeton University Press.

Numbers, R. L. (1976). *Prophetess of Health: A Study of Ellen G. White.* New York: Harper & Row.

Oakley, B. (2011). *Cold-Blooded Kindness: Neuroquirks of a Codependent Killer, or Just Give Me a Shot at Loving You, Dear, and Other Reflections on Helping That Hurts.* Amherst, NY: Prometheus.

Oakley, B., Knafo, A., Madhavan, G., & Wilson, D. S. (2011). *Pathological Altruism.* Oxford: Oxford University Press.

O'Brien, D. T. (2012). "Managing the Urban Commons: The Relative Influence of Individual and Social Incentives on the Treatment of Public Space." *Human Nature, 23*(4), 467–489. doi:10.1007/s12110-012-9156-6

O'Brien, D. T., Gallup, A. C., & Wilson, D. S. (2012). "Residential Mobility and Prosocial Development within a Single City." *American Journal of Community Psychology, 50*(1–2), 26–36. doi:10.1007/s10464-011-9468-4

O'Brien, D. T., & Kauffman, R. A. (2012). "Broken Windows and Low Adolescent Prosociality: Not Cause and Consequence, but Co-Symptoms of Low Collective Efficacy." *American Journal of Community Psychology.* doi:10.1007/s10464-012-9555-1.

O'Brien, D. T., Norton, C. C., Cohen, J., & Wilson, D. S. (2012). "Local Adaptation in Community Perception: How Background Impacts Judgments of Neighborhood Safety." *Environment and Behavior. 46*(2), 213–240. doi:10.1177/0013916512456844

O'Brien, D. T., & Wilson, D. S. (2011). "Community Perception: The Ability to Assess the Safety of Unfamiliar Neighborhoods and Respond Adaptively." *Journal of Personality and Social Psychology, 100*(4), 606–620. doi:10.1037/a0022803.

O'Conner, L. E., Berry, J. W., Lewis, T. B., & Stiver, D. J. (2011). Empathy-based Pathogenic Guilt, Pathological Altruism, and Psychopathology. In B. Oakley, A. Knafo, G. Madhavan, & D. S. Wilson (Eds.), *Pathological Altruism* (pp. 10–30). Oxford: Oxford University Press.

Okasha, S. (2006). *Evolution and the Levels of Selection.* Oxford: Oxford University Press.

Okasha, S. (2011a). "Précis of Evolution and the Levels of Selection." *Philosophy and Phenomenological Research, 82*(1), 212–220. doi:10.1111/j.1933-1592.2010.00470.x.

Okasha, S. (2011b). "Reply to Sober and Waters." *Philosophy and Phenomenological Research, 82*(1), 241–248. doi:10.1111/j.1933-1592.2010.00474.x

Okasha, S. (2014). "The Relation between Kin and Multilevel Selection: An Approach Using Causal Graphs." *British Journal for the Philosophy of Science,* in press.

Ostrom, E. (1990). *Governing the Commons: The Evolution of Institutions for Collective Action.* Cambridge: Cambridge University Press.

Ostrom, E. (2010). "Beyond Markets and States: Polycentric Governance of Complex Economic Systems." *American Economic Review, 100*, 1–33.

Pagel, M. (2012). *Wired for Culture: The Natural History of Human Cooperation.* New York: Allen Lane.

Pagel, M., & Mace, R. (2004). "The Cultural Wealth of Nations." *Nature, 428,* 275–278.

Passino, K. M., Seeley, T. D., & Visscher, P. K. (2007). "Swarm Cognition in Honey Bees." *Behavioral Ecology and Sociobiology, 62*(3), 401–414. doi:10.1007/s00265-007-0468-1

Pepper, J., Findlay, S. C., Kassen, R., Spencer, S., & Maley, C. (2009). "Cancer Research Meets Evolutionary Biology." *Evolutionary Applications, 2*, 62–70.

Pickett, K., & Wilkinson, J. B. (2009). *The Spirit Level: Why Greater Equality Makes Societies Stronger.* London: Bloomsbury Press.

Piersma, T., & van Gils, J. A. (2010). *The Flexible Phenotype: A Body-Centered Integration of Ecology, Physiology, and Behavior.* Oxford: Oxford University Press.

Pigliucci, M. (2001). *Phenotypic Plasticity: Beyond Nature and Nurture.* Baltimore, MD: Johns Hopkins University Press.

Piketty, T. (2014). *Capital in the Twenty-First Century.* Cambridge, MA: Belknap Press.

Popper, K. (1934). *The Logic of Scientific Discovery.* New York: Routledge.

Prins, H. H. T. (1995). *Ecology and Behaviour of the Africa Buffalo: Social Inequality and Decision Making.* Berlin: Springer.

Prinz, R. J., Sanders, M. R., Shapiro, C. J., Whitaker, D. J., & Lutzker, J. R. (2009). "Population-Based Prevention of Child Maltreatment: The U.S. Triple P Population Trial." *Prevention Science, 10*, 1–12.

Putnam, R. D. (2000). *Bowling Alone: The Collapse and Revival of American Community.* New York: Simon and Schuster.

Queller, D. C. (1991). "Group Selection and Kin Selection." *Trends in Ecology and Evolution., 6*(2), 64

Rand, A. (1957). *Atlas Shrugged.* New York: Random House.

Rand, A. (1961). *The Virtue of Selfishness.* New York: Signet.

Rawls, J. (1971). *A Theory of Justice.* Cambridge, MA: Harvard University Press.

Richerson, P. J., & Boyd, R. (1999). "Complex Societies: The Evolutionary Origins of a Crude Superorganism." *Human Nature, 10*, 253–290.

Richerson, P. J., & Boyd, R. (2005). *Not by Genes Alone: How Culture Transformed Human Evolution.* Chicago: University of Chicago Press.

Sampson, R. J. (2003). "The Neighborhood Context of Well-Being." *Perspectives in Biology and Medicine, 46*, S53–S64.

Sampson, R. J. (2004). "Neighborhood and Community: Collective Efficacy and Community Safety." *New Economy, 11*, 106–113.

Scott-Phillips, T. C., Dickins, T. E., & West, S. A. (2011). "Evolutionary Theory and the Ultimate-Proximate Distinction in the Human Behavioral Sciences." *Perspectives on Psychological Science, 6*, 38–47.

Seeley, T. D. (1995). *The Wisdom of the Hive.* Cambridge, MA: Harvard University Press.

Seeley, T. D. (2010). *Honeybee Democracy.* Princeton, NJ: Princeton University Press.

Seeley, T. D., & Buhrman, S. C. (1999). "Group Decision Making in Swarms of Honey Bees." *Behavioral Ecology and Sociobiology, 45*, 19–31.

Seeley, T. D., Visscher, P. K., Schlegel, T., Hogan, P. M., Franks, N. R., & Marshall, J. R. (2012). "Stop Signals Provide Cross Inhibition in Collective Decision Making by Honeybee Swarms." *Science, 335*(6064), 108–11. doi:10.1126/science.1210361

Simon, B. (2014). "Continuous-Time Models of Group Selection, and the Dynamical Insufficiency of Kin Selection Models." *Journal of Theoretical Biology, 349*, 22–31. doi:10.1016/j.jtbi.2014.01.030

Simon, B., Fletcher, J. A., & Doebeli, M. (2013). "Towards a General Theory of Group Selection." *Evolution: International Journal of Organic Evolution, 67*(6), 1561–1572. doi:10.1111/j.1558-5646.2012.01835.x

Smith, A. (1759). *The Theory of Moral Sentiments.* New York: Empire.

Smith, A. (1776). *The Wealth of Nations.* Hollywood, FL: Simon and Brown.

Smith, C. (2003). *Moral, Believing Animals.* Oxford: Oxford University Press.

Sober, E. (1984). *The Nature of Selection: Evolutionary Theory in Philosophical Focus.* Cambridge, MA: Bradford/MIT.

Sober, E. (2011). "Realism, Conventionalism, and Causal Decomposition in Units of Selection: Reflections on Samir Okasha's Evolution and the Levels of Selection." *Philosophy and Phenomenological Research, 82*(1), 221–231. doi:10.1111/j.1933-1592.2010.00471.x

Sober, E., & Wilson, D. S. (1998). *Unto Others: The Evolution and Psychology of Unselfish Behavior.* Cambridge, MA: Harvard University Press.

Sontag, C., Wilson, D. S., & Wilcox, R. S. (2006). "Social Foraging in *Bufo Americanus* Tadpoles." *Animal Behaviour, 72*(6), 1451–1456. doi:10.1016/j.anbehav.2006.05.006.

Stoelhorst, J. W., & Richerson, P. J. (2013). "A Naturalistic Theory of Economic Organizations." *Journal of Economic Behavior and Organization.* In press.

Stone, B. L. (2010). "The Current Evidence for Hayek's Cultural Group Selection Theory." *Libertarian Papers, 2.*

Strobel, A., Zimmermann, J., Schmitz, A., Reuter, M., Lis, S., Windmann, S., & Kirsch, P. (2011). "Beyond Revenge: Neural and Genetic Bases of Altruistic

Punishment." *NeuroImage*, *54*(1), 671–80. doi:10.1016/j.neuroimage.2010.07.051.

Thaler, R. H., & Sunstein, C. R. (2008). *Nudge: Improving Decisions about Health, Wealth, and Happiness*. New Haven, CT: Yale University Press.

Tinbergen, N. (1963). "On Aims and Methods of Ethology." *Zeitschrift Für Tierpsychologie*, *20*, 410–433.

Tocqueville, A. de. (1835). *Democracy in America*. New York: Penguin Classic.

Tomasello, M. (2009). *Why We Cooperate*. Boston: MIT Press.

Tomasello, M., Carpenter, J., Call, J., Behne, T., & Moll, H. (2005). "Understanding and Sharing Intentions: The Origins of Cultural Cognition." *Behavioral and Brain Sciences*, *28*, 675–735.

Traulsen, A. (2010). "Mathematics of Kin- and Group-Selection: Formally Equivalent?" *Evolution: International Journal of Organic Evolution*, *64*(2), 316–323. doi:10.1111/j.1558-5646.2009.00899.x

Turchin, P. (2005). *War and Peace and War*. Upper Saddle River, NJ: Pi Press.

Turchin, P. (2011). "Warfare and the Evolution of Social Complexity: A Multilevel Selection Approach." *Structure and Dynamics*, *4*(3). Article 2:1-37.

Tyler, E. B. (1871). *Primitive Culture*. New York: Brantano's.

van Veelen, M., García, J., Sabelis, M. W., & Egas, M. (2012). "Group Selection and Inclusive Fitness Are Not Equivalent: The Price Equation vs. Models and Statistics." *Journal of Theoretical Biology*, *299*, 64–80. doi:10.1016/j.jtbi.2011.07.025.

Veblen, T. (1898). "Why Is Economics Not an Evolutionary Science?" *Quarterly Journal of Economics*, *12*, 373–397.

Wade, N. (2009). *The Faith Instinct: How Religion Evolved and Why It Endures*. New York: Penguin.

Waters, K. C. (2011). "Okasha's Unintended Argument for Toolbox Theorizing." *Philosophy and Phenomenological Research*, *82*(1), 232–240. doi:10.1111/j.1933-1592.2010.00472.x

Wegner, D. M. (1986). Transactive Memory: A Contemporary Analysis of the Group Mind. In B. Mullen & G. R. Goethals (Eds.), *Theories of Group Behavior* (pp. 185-208). New York: Springer-Verlag.

West-Eberhard, M. J. (2003). *Developmental Plasticity and Evolution*. Oxford: Oxford University Press.

Widiger, T. A., & Presnall, J. R. (2011). Pathological Altruism and Personality Disorder. In B. Oakley, A. Knafo, G. Madhavan, & D. S. Wilson (Eds.), *Pathological Altruism* (pp. 85–93). Oxford: Oxford University Press.

Williams, G. C. (1966). *Adaptation and Natural Selection: A Critique of Some Current Evolutionary Thought*. Princeton, NJ: Princeton University Press.

Wilson, D. S. (1975). "A Theory of Group Selection." *Proceedings of the National Academy of Sciences*, *72*, 143–146.

Wilson, D. S. (1977). "Structured Demes and the Evolution of Group Advantageous Traits." *American Naturalist, 111*, 157–185.

Wilson, D. S. (1995). "Language as a Community of Interacting Belief Systems: A Case Study Involving Conduct toward Self and Others." *Biology and Philosophy, 10*, 77–97.

Wilson, D. S. (1998). "Adaptive Individual Differences within Single Populations." *Philosophical Transactions of the Royal Society of London, Series B, 353*, 199–205.

Wilson, D. S. (2000). Animal Movement as a Group-Level Adaptation. In S. Boinski & P. A. Garber (Eds.), *On the Move: How and Why Animals Travel in Groups* (pp. 238–258). Chicago: University of Chicago Press.

Wilson, D. S. (2002). *Darwin's Cathedral: Evolution, Religion and the Nature of Society.* Chicago: University of Chicago Press.

Wilson, D. S. (2005). "Testing Major Evolutionary Hypotheses about Religion with a Random Sample." *Human Nature, 16*, 382–409.

Wilson, D. S. (2007). *Evolution for Everyone: How Darwin's Theory Can Change the Way We Think about Our Lives.* New York: Delacorte.

Wilson, D. S. (2008). "Social Semantics: Toward a Genuine Pluralism in the Study of Social Behaviour." *Journal of Evolutionary Biology, 21*(1), 368–373. doi:10.1111/j.1420-9101.2007.01396.x

Wilson, D. S. (2009). Multilevel Selection and Major Transitions. In M. Pigliucci & G. B. Muller (Eds.), *Evolution: The Extended Synthesis* (pp. 81–94). Cambridge, MA: MIT Press.

Wilson, D. S. (2011). *The Neighborhood Project: Using Evolution to Improve My City, One Block at a Time.* New York: Little, Brown.

Wilson, D. S. (2012a). "A Tale of Two Classics." *New Scientist, 30*, 30–31.

Wilson, D. S. (2012b). "Human Cultures Are Primarily Adaptive at the Group Level." Social Evolution Forum. *Cliodynamics: The Journal of Theoretical and Mathematical History, 4*(1). http://escholarship.org/uc/item/05n4z9w8

Wilson, D. S., & Colwell, R. K. (1981). "The Evolution of Sex Ratio in Structured Demes." *Evolution, 35*, 882–897.

Wilson, D. S., & Csikszentmihalyi, M. (2007). Health and the Ecology of Altruism. In S. G. Post (Ed.), *The Science of Altruism and Health* (pp. 314–331). Oxford: Oxford University Press.

Wilson, D. S., & Gowdy, J. M. (2013). "Evolution as a General Theoretical Framework for Economics and Public Policy." *Journal of Economic Behavior and Organization, 90*, S3–S10. doi:10.1016/j.jebo.2012.12.008

Wilson, D. S., & Gowdy, J. M. (2015). "Human Ultrasociality and the Invisible Hand: Foundational Developments in Evolutionary Science Alter a Foundational Concept in Economics." Manuscript in progress.

Wilson, D. S., Hayes, S. C., Biglan, A., & Embry, D. (2014). "Evolving the Future:

Toward a Science of Intentional Change." *Behavioral and Brain Sciences*, 37, 395–460.

Wilson, D. S., Kauffman, R. A., & Purdy, M. S. (2011). "A Program for At-Risk High School Students Informed by Evolutionary Science." *PLoS ONE, 6* (11), e27826. doi:10.1371/journal.pone.0027826

Wilson, D. S., & O'Brien, D. T. (2009). "Evolutionary Theory and Cooperation in Everyday Life." In S. Levin (Ed.), *Games, Groups, and the Common Good*. Berlin: Springer. (pp. 155–163)

Wilson, D. S., O'Brien, D. T., & Sesma, A. (2009). "Human Prosociality from an Evolutionary Perspective: Variation and Correlations at a City-Wide Scale." *Evolution and Human Behavior, 30*(3), 190–200. doi:10.1016/j.evolhumbehav.2008.12.002.

Wilson, D. S., Ostrom, E., & Cox, M. E. (2013). "Generalizing the Core Design Principles for the Efficacy of Groups." *Journal of Economic Behavior and Organization, 90*, S21–S32. doi:10.1016/j.jebo.2012.12.010

Wilson, D. S., & Sober, E. (1989). "Reviving the Superorganism." *Journal of Theoretical Biology, 136*, 337–356.

Wilson, D. S., & Sober, E. (1994). "Reintroducing Group Selection to the Human Behavioral Sciences." *Behavioral and Brain Sciences, 17*, 585–654.

Wilson, D. S., & Sober, E. (2002). "Multilevel Selection, Pluralism, and All That (Precis of *Unto Others* and Reply to Commentaries)." *Philosophy and Phenomenological Research*, 65(3), 681–727.

Wilson, D. S., & Wilson, E. O. (2007). "Rethinking the Theoretical Foundation of Sociobiology." *Quarterly Review of Biology, 82*, 327–348.

Wilson, E. O. (1975). *Sociobiology: The New Synthesis*. Cambridge, MA: Harvard University Press .

Wilson, E. O. (2012). *The Social Conquest of Earth*. New York: Norton.

Wright, J. B. (2005). "Adam Smith and Greed." *Journal of Private Enterprise, 21*, 46–58.

Wright, R. (2009). *The Evolution of God*. New York: Little, Brown.

Zahn-Waxler, C., & Van Hulle, C. (2011). Empathy, Guilt, and Depression: When Caring for Others becomes Costly for Children. In B. Oakley, A. Knafo, G. Madhavan, & D. S. Wilson (Eds.), *Pathological Altruism* (pp. 321–344). Oxford: Oxford University Press.

# Index